T0296518

CAMBRIDGE MONOGRAPHS ON PHYSICS

GENERAL EDITORS

N. FEATHER, F.R.S.
Professor of Natural Philosophy in the University of Edinburgh

D. SHOENBERG, PH.D.
Fellow of Gonville and Caius College, Cambridge

SURFACE TENSION AND THE SPREADING OF LIQUIDS

SURFACE TENSION
AND THE
SPREADING OF LIQUIDS

BY

R. S. BURDON, D.Sc., F.Inst.P.

Reader in Physics, University of Adelaide

SECOND EDITION

CAMBRIDGE
AT THE UNIVERSITY PRESS
1949

CAMBRIDGE
UNIVERSITY PRESS

University Printing House, Cambridge CB2 8BS, United Kingdom

Published in the United States of America by Cambridge University Press, New York

Cambridge University Press is part of the University of Cambridge.

It furthers the University's mission by disseminating knowledge in the pursuit of education, learning and research at the highest international levels of excellence.

www.cambridge.org
Information on this title: www.cambridge.org/9781107666573

© Cambridge University Press 1949

This publication is in copyright. Subject to statutory exception and to the provisions of relevant collective licensing agreements, no reproduction of any part may take place without the written permission of Cambridge University Press.

First published in Cambridge Physical Tracts 1940
Second edition, revised and reset, and issued in
Cambridge Monographs on Physics 1949
First published 1949
First paperback edition 2014

A catalogue record for this publication is available from the British Library

ISBN 978-1-107-66657-3 Paperback

Cambridge University Press has no responsibility for the persistence or accuracy of URLs for external or third-party internet websites referred to in this publication, and does not guarantee that any content on such websites is, or will remain, accurate or appropriate.

GENERAL PREFACE

The Cambridge Physical Tracts, out of which this series of Monographs has developed, were planned and originally published in a period when book production was a fairly rapid process. Unfortunately, that is no longer so, and to meet the new situation a change of title and a slight change of emphasis have been decided on. The major aim of the series will still be the presentation of the results of recent research, but individual volumes will be somewhat more substantial, and more comprehensive in scope, than were the volumes of the older series. This will be true, in many cases, of new editions of the Tracts, as these are republished in the expanded series, and it will be true in most cases of the Monographs which have been written since the War or are still to be written.

The aim will be that the series as a whole shall remain representative of the entire field of pure physics, but it will occasion no surprise if, during the next few years, the subject of nuclear physics claims a large share of attention. Only in this way can justice be done to the enormous advances in this field of research over the War years.

N. F.
D. S.

CONTENTS

Author's Preface   *page* xiii

Chapter I. THE NATURE OF SURFACE FORCES   1

  1. Surface Tension and Surface Energy   1

  2. Origin of Surface Energy   2

  3. The Free Energy Relation   3

  4. Adsorption at Boundaries: Impurities   3

    (1) Composition of surface layer   3
    (2) Time-lag in reaching equilibrium in boundary
        layers   4
    (3) Attempts to prove surface adsorption in pure
        liquids   5
    (4) Measurement of adsorption at surfaces   6

  5. Recent Work: Maxima and Minima in γ-c Curves for
    Dilute Solutions   7

    (1) Inorganic salts   7
    (2) Capillary-active solutes   7

Chapter II. MEASUREMENT OF SURFACE TENSION   10

  1. General Remarks   10

    (1) Accuracy of measurement   10
    (2) Surface tension, curvature and pressure 10

  2. Methods of measuring Surface Tension   11

    (1) Capillary rise   12
    (2) Drop-weight or drop-number   13
    (3) Maximum bubble pressure   14
    (4) Measurement of large sessile drop   14

      (a) Infinitely large drop   15
      (b) Effect of curvature at summit 16
      (c) Finite diameter of drop   16

    (5) Oscillation of jets and drops   18
    (6) Capillary ripples   19

(7) Bond's flowing sheet method   *page* 20
(8) Adhesion of rings and plates   20
(9) Shape of pendent drop   20

3. Measurement of Small Differences in Surface
   Tension   22

   (1) Differential capillarimeter   22
   (2) Twin ring surface tensiometer   22

4. Interfacial Tension   22

Chapter III. THE SURFACE OF LIQUID METALS:
MERCURY   24

1. Metals in General   24

2. Mercury   25

   (1) The remarkable discordance in observations 25
   (2) Surface tension in vacuum and in gases   26
   (3) Capillary depression and meniscus volume   27
   (4) Temperature coefficient of γ 28
   (5) Gases and vapours on the surface of mercury 29

       (a) Effects on surface tension   29
       (b) Degree of persistence of adsorbed gases 30
       (c) Conclusions   31

3. Other Liquid Metals: Gallium   32

Chapter IV. SPREADING: GENERAL CONDITIONS   33

1. No Simple and Complete Theory of Spreading   33

2. Energy Conditions   34

3. Mechanism of Spreading on Liquids   36

4. Spreading on Solids   38

   (1) Energy condition: angle of contact   38
   (2) Mechanical conditions for spreading on solids 39

       (a) Effect of immobility of surface of a solid 39
       (b) The 'vapour-pressure' theory of spreading 40

   (3) Other factors in spreading on solids   41

       (a) Surface migration   41
       (b) Gravitational forces   41
       (c) Long-range molecular forces   41

Chapter V. SPREADING ON THE SURFACE OF
MERCURY   *page* 42

1. Phenomena of Spreading   42

2. Spreading of Water and Aqueous Solutions   43
   (1) Spreading to a limited area   43
   (2) The speed and mechanism of spreading   45

3. Spreading controlled by Electric Current across the
   Interface   48
   (1) Distilled water   48
   (2) Very dilute solutions of acids   48
   (3) Alkaline solutions   49
   (4) Comparison with capillary-electric phenomena 49

4. The Interfacial Tension of Mercury and Pure
   Water   50
   (1) Antonow's rule   50
   (2) Pure water as a solution of ions   51

Chapter VI. SPREADING ON WATER   52

1. Introduction   52

2. Non-spreading Liquids   52

3. Fatty Acids   53
   (1) Behaviour on the surface of water   53
   (2) Molecular areas from spreading   54

4. Measurements on Monomolecular Films on Water 55
   (1) The Langmuir-Adam balance   55
   (2) Types of film   57
      (*a*) Gaseous films   57
      (*b*) Close-packed or condensed films   58
      (*c*) Expanded films   58

5. Test of Homogeneity of Films   60
   (1) Optical methods   60
   (2) Method of surface-potential measurement 61

6. Solutions of Active Substances in Non-spreading Paraffins *page* 61

7. Nature of Interfacial Films and Monolayers 64
   (1) Effect of metallic ions 64
   (2) Complexes in monolayers 64

8. Multilayer Films on Solids 65

9. Paraffin caused to Spread by Irradiation 66

10. Spreading on a partially Contaminated Surface 68
    (1) Visible limit of a monolayer 68
    (2) Benzene on water 68
    (3) Behaviour of aniline on water 68

11. Action of Oil on Surface Waves on Water 69

12. Influence of Surface Layers on Evaporation 70

Chapter VII. LIQUIDS ON THE SURFACE OF SOLIDS 72

1. General Remarks 72

2. Lubrication 73
   (1) Complete lubrication: static friction absent 73
   (2) Boundary lubrication: static friction present 74
   (3) Recent work 76

3. Angle of Contact 76
   (1) Its reality and importance 76
   (2) Hysteresis of the contact angle 77
   (3) Measurement of contact angles 78
       (*a*) Plate method 78
       (*b*) Contact angles of fibres 79
       (*c*) Contact of bubbles 79
       (*d*) Contact angle in powders: wetting pressure 80
   (4) Surface structure and contact angle 81
   (5) Contact angle in liquids: Antonow's rule 81

4. Flotation of Minerals 82
   (1) Introduction 82
   (2) Principles of flotation 83

(3) Flotation reagents   *page* 84

    (*a*) Frothers   85
    (*b*) Collectors   85
    (*c*) Activators   85
    (*d*) Depressants   86

5. Detergent Action   86

6. Other Technical Applications: Wetting Agents   88

    (1) Wetting of metals by liquid metals   88
    (2) Wetting agents   88

Index   90

Plates facing pages: I 46; II 62; III 63

AUTHOR'S PREFACE

This tract deals with matters that have been subjects of experiment and discussion for at least a century. This makes it difficult to know what should be left out from an account written to enable readers, not engaged in similar work, rapidly to acquire an idea of the present position of the subject. A comprehensive list of references would occupy almost as much space as the text and those given here are designed merely to provide a starting-point for anyone wishing to pursue a section in detail.

The appearance within a few months of a new edition of Adam's *Physics and Chemistry of Surfaces*, the Report of the Leather Trade Chemists' *Symposium on 'Wetting and Detergency'*, and Wark's monograph on *The Flotation of Minerals*, has made available up-to-date accounts of most sections of the subject from both scientific and technical aspects. To these as well as to the texts of Rideal and Freundlich the author's indebtedness is gratefully acknowledged.

It has been assumed that the reader does not want detailed analyses or descriptions of experimental methods. Emphasis has been placed on things not in the text-books and where uncertainty exists, or progress is being made. The earlier part of the book presents the case that after a century of work we do not know the surface tension of mercury nor even exactly why it is difficult to determine this quantity, and that few have any ideas as to what methods of measuring surface tension give reliable results.

The general conditions affecting the spreading of liquids are discussed and some account is given of various investigations based on the phenomena of spreading. By far the greater part of the work on spreading is now being carried out under the auspices of technical and industrial organizations. The last chapter, which deals with one or two of the technical applications of spreading, could be extended almost indefinitely but without introducing any new principles of physics.

To the Physical Society and the Faraday Society are due my thanks for permission to use material published in their journals, also to Mr J. S. Anderson for drawing the figures and to Mr B. Worthley for reading the manuscript.

R. S. BURDON

Department of Physics
University of Adelaide

DECEMBER 1939

PREFACE TO THE SECOND EDITION

No attempt has been made to do more than extend the tract by including some account of work published in recent years, keeping the original aim in view.

The writer would thank his colleagues for their assistance and particularly Messrs G. J. Aitchison and G. J. Burdon for help with manuscript and figures.

R. S. BURDON

Adelaide

JANUARY 1949

CHAPTER I

THE NATURE OF SURFACE FORCES

1. Surface Tension and Surface Energy

The idea of a tension in the free surface of a liquid is familiar as an explanation of the tendency of a liquid surface to assume the form having a minimum area, as shown in the shape of a bubble or a drop of liquid. Though many phenomena lend a semblance of reality to this idea of a tension in the superficial layers of a liquid, a surface film possesses no modulus of elasticity, for a force that causes any expansion of the film will, if maintained, cause the film to expand indefinitely or to break. There is thus no 'elastic' film, the stretching is a creation of new surface and, as we shall see presently, the tendency to contract is merely a special case of the general principle that potential (or free) energy tends toward a minimum value.

If a soap film be thinned by evaporation, practically no diminution in its strength is observed so long as the film persists, a fact which shows that the phenomenon of surface tension is due to a layer not more than a few molecules in thickness at the most.

The surface tension, (γ), of a liquid is expressed in dynes per centimetre, and if a strip of surface one centimetre wide is extended through an additional centimetre of length, it is clear that γ ergs of work have been performed and also that one sq.cm. of new surface has been formed. Moreover, if the surface be allowed to contract reversibly, while its area decreases by one sq.cm., it will perform γ ergs of work. Thus, by analogy with other reversible processes, we see that the surface tension of a liquid is numerically equal to the free energy of its surface per unit area.

From many points of view it is desirable to discuss phenomena occurring at surfaces in terms of free energy rather than a tension in the surface, but the familiar expression 'surface tension' will probably remain.

[The units *dyne per cm.* and *erg per sq.cm.* have the same dimensions, but the mixed expression 'surface tension in ergs/cm.²' which is used occasionally is obviously wrong.]

2. Origin of Surface Energy

A molecule in a body of liquid is subject to forces due to its neighbours. These forces will fluctuate owing to a molecular agitation, but averaged over a finite period of time their effect will be zero, so that a molecule can move about in the liquid without doing any work against these forces. In coming to the free surface of the liquid, however, the molecule ultimately must move into the surface against unbalanced forces, owing to the fact that it is no longer surrounded symmetrically by other molecules. The direction of these forces which act on molecules forming the surface layer is essentially normal to the free surface of the liquid.

Thus molecules in the boundary layer, whether a free surface or the interface between two fluids, possess energy because work has been done in bringing them into the surface. This additional energy due to their superficial position may be regarded as surface energy.

This conception applies to both solids and liquids and it meets or rather avoids the objection which is still met on occasion, namely that a solid cannot have a surface tension. Molecules in the superficial layers of a solid, whether amorphous or crystalline, possess energy in virtue of their position and arrangement, and this conception of surface energy is as fundamental in discussing phenomena at the surface of solids as it is in the case of liquids. The wetting of solids as in lubrication, painting, soldering, washing and dyeing of fabrics or the separation of minerals by flotation is always determined by energy relations at surfaces. Even the loss of energy by magnetic hysteresis in iron is largely determined by the size of the crystal grain and hence by the proportion of the molecules that are in boundary layers and not in the crystal continuum.

Many attempts have been made to account quantitatively for surface energy in terms of various assumed laws of force between molecules. Lennard-Jones and Corner (*Trans. Far. Soc.* **36**, 1156, 1940) briefly review this work and show how modern views of the nature of intermolecular forces may be applied to give a theoretical basis to surface phenomena.

3. The Free Energy Relation

The familiar relation connecting free energy, heat of reaction, and temperature, viz.

$$F = H + T\frac{\partial F}{\partial T},$$

when applied to surface formation becomes

$$\gamma = H + T\frac{\partial \gamma}{\partial T}.$$

For all liquids γ decreases with rising temperature since it must vanish at the critical point [Ch. II, 2 (1)] so that $\partial\gamma/\partial T$ is negative and $\gamma < H$ for all liquids. Thus, H, the energy involved in forming a sq.cm. of new surface is always greater than γ, the work done in stretching the surface. The balance, $H - \gamma$, is supplied by a cooling of the neighbouring liquid when the surface is enlarged, and this quantity is sometimes called the latent heat of surface formation.

[Positive temperature-coefficient of surface tension:

Bircumshaw (*Phil. Mag.* **12**, 596, 1931) found some evidence indicating that for mercury, just above its freezing point, $d\gamma/dT$ was positive. This may possibly be due to some degree of organization persisting in the molecules above the melting point and giving rise to more than one molecular species [4 (1)].

Earlier workers have reported a positive temperature coefficient over some range of temperature for cadmium and also for iron and copper, but experimental difficulties make it at least doubtful if the case has been proved for these metals in a perfectly pure state. (Cf. Adam, *Physics and Chemistry of Surfaces*, p. 164, 1941.)]

4. Adsorption at Boundaries: Impurities

(1) *Composition of surface layer*. When a liquid contains more than one molecular species equilibrium will be reached only when the surface layers comprise those molecules which make the free energy of the surface a minimum. There will thus be formed a surface phase differing in composition from the bulk of the liquid. This difference may be of any degree, ranging from the case where a single layer of closely-packed molecules is adsorbed at the

boundary, such as a monomolecular layer of insoluble fatty acid on water (Ch. VI), down to the case where there is near the surface merely a slight excess of one kind of molecule, determined by the balance between the forces tending to promote adsorption and those causing diffusion.

(2) *Time-lag in reaching equilibrium in boundary layers.* A fall in surface tension due to the adsorption of a surface-active solute into the surface of a solution has been known since Rayleigh, nearly sixty years ago, found that the surface tension of a 0·025% solution of sodium oleate was less by 30% when the surface was at rest than when freshly formed and flowing. Estimates based on the theory of diffusion indicate that molecules to be adsorbed should reach the surface in a period usually too brief to be measured, whereas comparatively recent work shows that a surface, even though most carefully guarded against external contamination, may continue to fall in surface tension for hours or days before reaching its final value [e.g. Adam and Shute (*Trans. Far. Soc.* **34**, 758, 1938), McBain and Sharp (*Jour. Am. Chem. Soc.* **63**, 1422, 1941), Alexander (*Trans. Far. Soc.* **37**, 15, 1941), Alexander and Rideal (*Nature,* **155**, 18, 1945) etc.]. McBain and Sharp used the film balance (Ch. II) to show that the undisturbed surface of a solution of one gram per litre of hydrocinnamic acid fell progressively in surface tension below the value for a freshly swept surface. The difference was 0·4 dyne/cm. after 30 minutes and 1·5 dynes/cm. after 24 hours.

Adam and Shute, observing sessile bubbles (Ch. II) followed the fall in surface tension of dilute solutions of paraffin chain salts through a range of 40 dynes/cm. A 0·001N solution continued to fall in surface tension for a week or more. The fact that a fresh bubble would go through the same course showed that the effect was due to ageing of the surface and not to a change in the bulk of the dilute solution. The time required to attain equilibrium in the surface is much less at higher concentrations, and also if inorganic salts are present in solution.

The reason why the time required to reach final equilibrium may be a million times larger than that needed for diffusion to bring the molecules into the surface is not clear. In order to be adsorbed at the surface, molecules reaching there must be unassociated and perhaps orientated in a definite sense, and further, the final process of adsorption may be one requiring large activation energy as suggested by Ward and Tordai (*Nature*, **154**, 146, 1944). This idea of an energy barrier to adsorption at the surface is supported by the fact that many surfaces which show this time-lag also possess some of the properties of an insoluble surface layer, such as the power to resist compression by a floating barrier in the film trough (Ch. VI). Alexander and Rideal found that fatty acid molecules were adsorbed at an interface much more rapidly from a benzene solution than from heavy hydrocarbons, and suggested that the time for the surface to attain equilibrium gave a method of measuring the degree of dissociation of polar compounds in hydrocarbons.

Ross (*Jour. Phys. Chem.* **49**, 377, 1945) investigated the variation with time of the surface tension of dilute solutions of saponin. He regards the initial rapid fall from 72 to 62 dynes/cm. as due to adsorption, and the further slow fall, extending over an hour or more, as the result of some autocatalytic reaction between the adsorbed molecules and others coming into the surface from the underlying solution.

(3) *Attempts to prove surface adsorption in pure liquids.* Much effort has been devoted to attempts to show a time-lag in attaining equilibrium in the case of pure liquids, particularly water and mercury. It has been thought that the process of association might give rise to molecular groups which differed in surface energy and the surface having minimum energy might then take a measurable time to be formed. Hiss and also Schmidt and Steyer (*Ann. Physik*, **79**, 442, 1926) have endeavoured to trace the fall in surface tension of water during the first thousandth of a second after the surface is formed. Higher initial values were obtained for the surface tension, but the experimental difficulties of their method are such that there must remain some doubt about the result. Their work, as well as that of Bohr (*Phil. Trans.* A, **209**, 281, 1909), indicates that less than 0·01 second after its formation the surface of water has reached a state of constant surface energy. More recently Bond and Puls (*Phil. Mag.* **24**, 864, 1937) have come to the conclusion that equilibrium is reached much sooner than this in the case of pure liquids.

In the case of mercury, any fall in surface tension with time that has been observed is almost certainly due to the adsorption of impurity, a minute trace of which may profoundly affect the surface tension. The avoidance of such impurities forms one of the chief difficulties in the measurement of surface tension, particularly in the case of metals (Ch. III).

(4) *Measurement of adsorption at surfaces.* When selective adsorption from a vapour or solution occurs at a surface, the Gibbs adsorption equation, based on thermodynamical reasoning, indicates that the excess adsorbed at the surface is proportional to c, the concentration of the active component, and to $d\gamma/dc$, the slope of the γ-c curve.

Satisfactory confirmation of Gibbs's equation has been difficult owing to the problem of measuring the surface excess, particularly under conditions when equilibrium has been established between the forces of adsorption and diffusion. McBain and Humphrey (*Jour. Phys. Chem.* **36**, 300, 1932) devised a method by which a fast-moving blade was caused to skim a very thin layer from the surface of the liquid at rest. Analysis of the surface portion and the bulk of the liquid gave results in agreement with the Gibbs equation. Earlier measurements on flowing surfaces gave widely varying results. McBain, Mills and Ford (*Trans. Far. Soc.* **36**, 930, 1940) have now replaced the microtome by a method of sweeping the surface layer from the solution into a cell in the path of one beam of light in a Rayleigh refractometer. When this surface layer is dissolved in the liquid in the cell the concentration is found from the change in optical path through the cell. The new method has confirmed results obtained by the microtome method and has shown the time-lag of a surface in reaching adsorption equilibrium.

The converse process of accepting the adsorption equation and using it to calculate the surface excess from measurements on γ and c has been used by various workers. Here also it is necessary to be certain that γ is measured after equilibrium has been attained in the surface, that is to say, γ must be measured by a 'static' method (Ch. II, 2). It would appear however that the process of adsorption is too complex to be discussed satisfactorily in terms of the simplest form of the Gibbs equation.

5. Recent Work: Maxima and Minima in the γ-c Curves for Dilute Solutions

(1) *Inorganic salts.* The general effect of inorganic salts in producing a slight increase in γ for water is well known. There is a negative adsorption, the surface layer being pictured as practically pure water, and the increase in γ is nearly proportional to the concentration of ions in the bulk of the solution. However, Jones and Ray (*Jour. Am. Chem. Soc.* **59**, 187, 1937), using a method of great sensitivity, found that for a concentration of less than 0·001N a solution of KCl had a lower surface tension than pure water. The slight lowering, of the order of 0·02 dyne/cm., means a region in the γ-c curve where $d\gamma/dc=0$ and hence where there should be no adsorption according to the Gibbs equation. Yet the surface tension is lowered. Langmuir (*Science*, **88**, 430, 1938) suggested that the effect observed by Jones and Ray could be satisfactorily explained as due to a layer of water, held bound to the surface of the quartz capillary tube in their apparatus [Ch. II, 3 (1)]. The thickness of this bound layer of water molecules depends on the concentration of solute. Any change in the layer would change the effective diameter of the capillary tube and thus alter the capillary rise. Langmuir calculated that the effect would be of the same order as that observed by Jones and Ray, but two years later Dole and Swartout (*Jour. Am. Chem. Soc.* **62**, 3039, 1940) announced that they had obtained results substantially similar to those of Jones and Ray with KCl, using a differential ring tensiometer. The depression of γ for water by some inorganic salts in extreme dilution may be regarded as still in some doubt.

(2) *Capillary-active solutes.* Fig. 1 indicates the form of γ-c curves for lauryl sulphonic acid obtained by McBain and his collaborators (*Report on Progress in Physics*, **5**, 30, 1938). Here the reduction is more than 40 dynes/cm. at a concentration of 0·005N, followed by a rise of 12 dynes/cm. at about 0·01N, and then a further fall at higher concentrations. The Gibbs equation would thus appear to indicate two regions where $d\gamma/dc=0$, and where the surface composition is identical with that of the bulk of the liquid, though the surface energy in these regions is less than half that of pure water. Moreover, the slope of the curve between these regions

indicates strong *negative* adsorption in spite of the low value of the surface tension.

As might be expected, direct measurements by methods indicated earlier [4 (4)] do in fact show positive adsorption throughout, irrespective of the sign of $d\gamma/dc$.

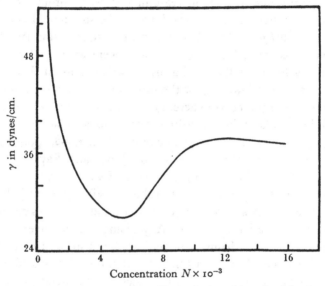

Fig. 1. Shape of γ-c curve for a dilute solution of lauryl sulphonic acid

McBain would account for the behaviour of these dilute solutions in terms of processes occurring in a surface layer that is very deep compared with molecular dimensions. He suggests that the Gibbs equation should be modified to include terms depending on the orientation of molecules, the formation of submerged double layers, and possibly other factors. Alexander (*Trans. Far. Soc.* **38**, 54, 1942) calculates the adsorption from pressure-area curves (Ch. VI) determined on the surfaces, assuming the pressure to be due entirely to adsorption into a monomolecular surface layer. His results agree with the direct measurements of McBain and he supports the earlier view that there is no need to postulate anything beyond a monolayer adsorbed to the surface of solutions. The apparent breakdown of the Gibbs equation is held to be due to the formation of molecules into micelles,* a process which has been shown to set in at concentrations near to those giving the minimum on the γ-c curve. It is found that the surface rapidly acquires its final equilibrium if the concentration is appreciably greater than that at which micelle formation sets in.

Reichenberg has recently discussed the various opinions regarding these anomalies (*Trans. Far. Soc.* **43**, 467, 1947). He concludes that minima in the γ-c curves cannot occur if both water and solute are perfectly pure. Minute traces of foreign ions, however, would explain the existence of a minimum in the curve. Reichenberg quotes experimental evidence of the anomaly disappearing when specially pure water was used as the solvent, instead of ordinary distilled water. Further experimental work with specially purified materials will be needed to settle the question.

* Micelle: a group of molecules, plane or possibly spherical in shape, with hydrocarbon chains directed inward and active ends outward.

CHAPTER II

MEASUREMENT OF SURFACE TENSION

1. General Remarks

(1) *Accuracy of measurement.* Many phenomena depending on surface tension have been described and a surprising number of them have been used to measure surface tension and its variations. Not infrequently considerable analysis is involved in deriving the formula by which the value of γ is to be calculated from the observations, and it is often safest to regard the value obtained for the surface tension merely as a measure of the success of the mathematical analysis. It was stated a few years ago that in no method except that of capillary rise was the theory sufficiently exact that only experimental errors remained. The position is a little better now, but it is probably fortunate that in most investigations on surface phenomena it has been sufficient to measure relative values or variations in γ rather than its actual value. It is easy to detect a variation of one part in 1000 in surface tension, but for no liquids except water and benzene have the absolute values of γ been agreed upon to within one-tenth of 1%. Since the measurement of surface tensions is often incidental to some other problem, it is perhaps well to stress the fact that no method should be accepted as satisfactory merely because it gives a consistent set of readings; both theory and experimental method need to be examined critically with a view to the particular problem on hand.

Recent developments have been chiefly in differential methods, capable of measuring small differences in γ or of following the 'ageing' of surfaces. Some of these methods are now very precise.

(2) *Surface tension, curvature and pressure.* When the surface of a liquid is curved, the pressure on the inside exceeds that on the outside by the amount

$$\gamma\left(\frac{1}{R_1}+\frac{1}{R_2}\right),$$

where R_1 and R_2 are the principal radii of curvature of the surface. Hence pressure inside a spherical bubble exceeds that on the outside by $4\gamma/R$, i.e. by $2\gamma/R$ for each surface, the inner and the outer.

For any continuous area of film over which the pressure difference between inside and outside is constant, the total curvature of the film must be the same at all points. In particular, if the pressure is the same on each side, the total curvature of the film must be zero everywhere. For instance, if we blow a cylindrical

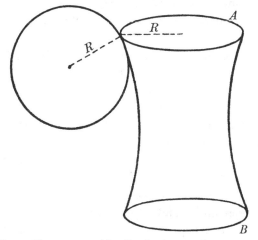

Fig. 2. Shape assumed by film having equal pressures on either side—total curvature everywhere zero

bubble (as can be done with the aid of two cylindrical tubes) and then allow the inside to come to atmospheric pressure, the film at once takes the shape indicated in Fig. 2. This is a figure of zero total curvature, since at any point the curvature of the film in the plane of the paper is equal and opposite to its curvature in a plane perpendicular to the paper.

The relation between curvature, pressure and surface tension is involved in most methods of measuring γ, but only in the method of capillary rise [2 (1)] can the relation given above be applied directly to give even an approximate value for surface tension.

2. Methods of measuring Surface Tension

In the following paragraphs little more than brief comments on the methods will be given, except where the points of interest are new or are not generally to be found in texts.

[References to many of the papers on the development of methods of measuring γ are given in Freundlich, *Colloid and Capillary Chemistry*, or Adam, *Physics and Chemistry of Surfaces* (1941).]

Methods are classed as *static* where the measurements are on a surface at rest, and *dynamic* where the surface is moving or where a fresh surface is continually being formed. Only where there is a time-lag in reaching equilibrium in the surface (Ch. I, 4) should there be any difference between the results from static and dynamic methods. The differences found are generally due to experimental error and lack of a complete theory of the method.

Evidently some methods cannot be classed definitely as either static or dynamic [e.g. (2), (3), (6)].

(1) *Capillary rise.* If the meniscus (Fig. 3) is part of a spherical surface of radius r, then from the relation between curvature and pressure we have

$$2\gamma/r = \rho g h.$$

If the tube is narrow and the angle of contact between liquid and tube is zero (liquid wets the tube), then r in the equation is very nearly equal to the radius of the tube. For the most precise work however it is necessary to correct for the shape of the meniscus (Rayleigh, *Proc. Roy. Soc.* A, **92**, 184, 1916). In particular, Porter has pointed out (*Nature*,

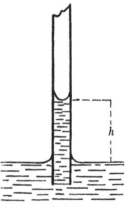

Fig. 3 Rise of liquid in a capillary tube

130, 929, 1932) that the current belief that the surface tension of a liquid falls practically to zero some distance below the critical temperature is probably wrong. Measurements near the critical point have generally been made by the method of capillary rise and it has not been recognized that a tube which is 'narrow' for large values of h will function as a 'wide' tube, requiring the meniscus correction, when h becomes very small as the critical point is approached.

The standard values for surface tension given in *International Critical Tables*, Vol. IV, are based on determinations made by this method. These values are 72.75 ± 0.05 dynes/cm. for water and

28·88 ± 0·03 for benzene, each under atmospheric pressure at 20° C.; and it may be said that in practice the usefulness of any method of measuring γ is finally judged by the degree to which it can reproduce these values.

The method of capillary rise is static, and obviously it can be adapted for use either in vacuum or with the liquid subjected to high pressure by gases. For the most precise work ρ must be replaced by the difference in density of the liquid and the air or other gas present.

[For a given liquid, using different tubes, the product $rh(= 2\gamma/\rho g)$ is obviously constant, and this expression is the 'capillary constant' frequently used in calculations by earlier writers on surface tension.]

A modification of the method of capillary rise, for use where only very small amounts of liquid are available, has been described by Ferguson (*Endeavour*, **2**, 34, 1943). A capillary tube having a flat end is mounted horizontally, and contains a drop of the liquid. Gas pressure is applied to force the drop to the end of the tube. When the liquid surface at the open end of the capillary tube is plane the applied pressure must be $2\gamma/r$. A manometer measures this pressure and examination of the reflection of a lamp-filament enables one to judge accurately when the liquid film and the flat end of the capillary tube are in one plane.

(2) *Drop-weight or drop-number.* These are perhaps the most widely-used methods; they can be used for determinations at any temperature or pressure and also for measuring interfacial tensions. There is, however, no simple formula which gives the maximum weight of a drop formed slowly at the tip of a capillary tube, in terms of the radius of the tube and the surface tension of the liquid, nor is it safe to assume that for a given tube the weights of drops will be strictly proportional to surface tension.

The experimental work of Harkins and Brown (*Jour. Am. Chem. Soc.* **41**, 499, 1919) established this method on a sound basis. Their expression for the weight W of a drop from a tube of radius r is

$$W = 2\pi r\gamma F\ (r/V^{\frac{1}{3}}),$$

where V is the volume of the drop. Values of F are now given in *International Critical Tables*, thus making the method readily available.

This method cannot be classified very definitely as either static or dynamic. For precise work it is arranged that the final stages in the formation and detachment of the drops take place very slowly, though the variation in weight due to the more rapid dropping rate is not very great. Where surface equilibrium is attained fairly slowly (as in sodium oleate solution) it is possible to follow the fall in surface tension with time by making observations using different rates of dropping.

Relative values of γ are quickly obtained, but only with doubtful accuracy, by counting the number of drops from a standard tip that is necessary to deliver definite amounts of the liquids.

(3) *Maximum bubble pressure.* If a cylindrical tube dips below the surface of a liquid and bubbles are formed slowly at the tip, then the maximum pressure reached (beyond hydrostatic pressure due to the depth of the tip below the free surface) depends on the surface tension of the liquid.

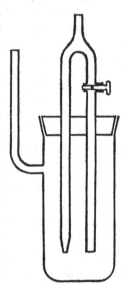

By using two tubes of different bore mounted to be accurately on the same horizontal level (Fig. 4) and finding the pressure to release bubbles from each tube in turn, the surface tension of a liquid may be rapidly determined. An empirical formula may be used, the constant for the instrument being found by using a standard liquid such as benzene (Sugden, *The Parachor and Valency*, p. 208). The method has come into considerable use of late. Using a pure neutral gas to blow the bubbles, surface contamination is eliminated, since each bubble forms a fresh surface in the body of the liquid. It is perhaps the best method for use with liquid metals, for which γ measured against a neutral gas is generally very close to the value found for the metal in vacuum. (Cf. Ch. III.)

Fig. 4. Apparatus for bubble-pressure method

(4) *Measurement of large sessile drop.* Some workers have determined γ from the relation between curvature and pressure [1 (2)], but here we shall give an outline of the method used more generally, where the measurements are simply those of the height and dia-

meter of the drop. Measurements of large drops have resulted in some of the most remarkable discrepancies in the literature, and as the standard text-books still do not contain a clear presentation of the method, a slightly fuller account here may serve a useful purpose, if only to illustrate the confusion that may arise from accepting a formula without investigation.

When a large circular drop of liquid is formed on a plane or concave surface (which it does not wet) the form of the drop depends on the surface tension and specific gravity of the liquid.

(*a*) *Infinitely large drop.* For an *infinitely* large drop there would be no curvature at the summit and any portion of the circumference could be regarded as straight. In such a case the

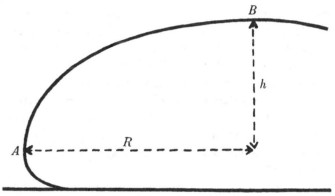

Fig. 5. Profile of a large drop of liquid resting
on a horizontal plane

surface tension of a strip of unit width at *B* (Fig. 5) may be regarded as supporting the total horizontal thrust of the liquid down to *A* where the film is vertical. Thus

$$\gamma = \rho g h \times h/2 = \tfrac{1}{2}\rho g h^2. \qquad (1)$$

Over half a century ago Worthington (*Phil. Mag.* **20**, 51, 1885) pointed out that this formula might be 10% or more in error for drops of the size measured by Quincke, who had used the formula given above. Since then some workers have used this formula, and others have employed one or other of the corrections to the formula, with the result that workers making measurements on mercury have published almost identical values for the surface tension of mercury though a scrutiny of their papers shows that the actual readings from which the calculations were made differ by about one part in ten.

Probably much of the confusion, bringing this method into some disrepute, arose from ignorance of the rather interesting fact that a large drop formed on a horizontal surface is shallower than a small one. If we run liquid from a burette on to a horizontal surface (which is not wetted) the drop increases in depth as it grows till a diameter of about 3 cm. is reached and thereafter the drop decreases slowly in depth as more liquid is added. The result is that the simple formula (1, above) is valid for drops of *over one metre* in diameter (which no one has used) and also for drops having a particular diameter slightly less than 2 cm. Some workers, apparently fortuitously, have used drops of very nearly the latter size.

(*b*) *Effect of curvature at summit.* If the surface of the drop at *B* has a radius of curvature *r*, we have seen [1 (2)] that this involves a hydrostatic pressure of $2\gamma/r$ which will act at every point in the drop, in addition to the pressure due to the head of mercury at that point. Thus *h* is less than it would be if the surface were truly flat at *B*. Worthington's analysis, as well as measurements made more recently by Gibson (*Proc. Roy. Soc. S. Aust.* **56**, 51, 1932), indicates that for drops having a diameter greater than 4 cm. this effect is negligible.

(*c*) *Finite diameter of drop.* The simple formula (1) assumes that the surface of the drop is only curved in the vertical plane, but of course there is curvature in any horizontal plane through the drop, the radius of curvature being *R* at the point *A*. Pressure due to this curvature in the horizontal plane assists in supporting the drop, and tends to make *h* greater for a smaller than for a larger drop [2 (4) (*a*)]. Worthington's evaluation of the total effect of this curvature leads to a formula (for $R > 2$ cm.) which may be written

$$\gamma = \frac{1}{2}\rho g h^2 \frac{1 \cdot 641 R}{1 \cdot 641 R + h}. \tag{2}$$

The correction factor is still appreciable for drops 15 cm. in diameter, but it involves only the measurement of the maximum diameter, which can be done well enough with an ordinary scale.

The position of A (Fig. 5) is easily and precisely found by setting the microscope on the image at A of a small lamp which is situated some distance away on the same horizontal plane as A. Parallel light may be used to define the summit B, and h can then be measured with a travelling microscope. The correct definition of B is not so easily attained as that of A and the method of defining it should be carefully checked. (*Nature*, **128**, 456, 1931.) It is probable that some of the discrepancies, particularly in the case of the highly reflecting surface of mercury, have been due to error in defining the summit of the drop, since a perfect reflector is of course invisible and can be located only from the appearance of objects reflected in the surface.

In a recent paper Kemball (*Trans. Far. Soc.* **42**, 526, 1946) describes what is probably the most satisfactory method of locating B when the drop is enclosed. A tungsten wire sealed through the glass carries a spring which supports an iron cylinder having a tungsten pointer at the bottom. An external magnet acting on the iron enables the tungsten point to be brought close to the surface which can then be located by the point and its re-flection. Kemball also shows that a serious error in h can arise from refraction if the glass window through which measurements are made has not truly parallel sides.

Gibson (*Proc. Roy. Soc. S. Aust.* **56**, 51, 1932) measured h for drops of distilled water formed in circular depressions freshly turned in paraffin wax, in order to test the formula (2). His results are shown in Fig. 6. The upper curve shows his values for γ cal-culated from the formula $\gamma = \frac{1}{2}\rho gh^2$ and the lower curve those obtained by using the correction factor, $1\cdot64R/(1\cdot64R+h)$. The curves show clearly

(i) that the simple formula is in error by several per cent even for drops as large as 12 cm. in diameter;

(ii) that the correction for horizontal curvature alone is suffi-cient for drops more than 4 cm. in diameter;

(iii) that for drops less than 4 cm. in diameter the curvature of the summit has an increasing effect, and for a particular diameter below 2 cm. this effect reduces h to the value it would have for an infinitely large drop. The curves indicate that the effect of curvature at the summit of the drop is varying rapidly with the diameter for drops of this size, so that in practice a diameter greater than 4 cm. should be used, when curvature at the summit can be neglected.

This method is clearly a static method. It has the advantage that measurements are taken only on parts of the liquid which are not in contact with any other surface. Suitably enclosed, the drop may be measured either in vacuum or under pressure due to gases.

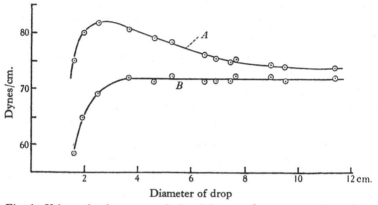

Fig. 6. Values of γ for water, calculated from measurements on large drops. A, using formula (1); B, using formula (2)

Kemball and Rideal have used the sessile drop in their studies of the adsorption of vapours on mercury (*Proc. Roy. Soc.* A, **187**, 53, 1946). The large drop is useful for following slow changes in surface tension and the method has been applied to the measurement of interfacial tension (when of course ρ would be replaced by the difference in densities of the two fluids).

Observation of sessile bubbles formed in a liquid beneath a slightly concave glass surface enables the changes in surface tension due to ageing of a liquid-gas interface to be followed. The bubbles behave essentially as inverted sessile drops.

(5) *Oscillation of jets and drops.* Falling drops, as well as jets from non-circular orifices, will oscillate about their stable form at rates dependent on the surface tension of the liquid. Lord Rayleigh worked out the theory of such oscillations, and considerable attention was devoted to methods based on these phenomena, in efforts to show the difference between dynamic and static surface tension.

Bohr (*Phil. Trans.* A, **209**, 281, 1909) used the method of the oscillating jet after making an analysis of the effects of viscosity, finite amplitude of oscillation, and the presence of air. Bohr's individual measurements for water lay between 72·98 and 73·41 dynes/cm., indicating that in his hands it was a method of precision. His results showed that a moving water surface a small fraction of a second after its formation has almost exactly the same surface tension as a surface at rest.

(6) *Capillary ripples.* For the velocity of waves on deep water Lord Kelvin derived the formula

$$v^2 = \frac{g\lambda}{2\pi} + \frac{2\pi\gamma}{\lambda\rho},$$

whence γ can be found if the wave-length, λ, and either the velocity or the frequency of the waves is known. For ripples such as those set up by pointers carried on the prongs of a tuning-fork the first term is almost negligible. The wave-length may be determined either by the use of stroboscopic illumination with waves set up by a single vibrating point, or from the interference pattern produced by ripples from two vibrating points. In spite of the stretching and contracting of the surface this is not a dynamic method, since it has been used to follow the fall of surface tension of mercury with time of exposure to the air. The method has been used recently by Brown (*Proc. Phys. Soc.* **48**, 312, 1936), who shows clearly that the amount of expansion and contraction occurring for the ripples used ($\lambda = 0·2$ cm.) is too small for the surface to behave otherwise in respect to adsorption than it would if at rest.

Satterly and Strachan (*Trans. Roy. Soc. Canada*, **29**, 109, 1935) have made measurements on the stationary ripples which may be observed on a circular jet of liquid when it falls vertically on to a horizontal surface. A difficulty in this experiment is to find the actual velocity in the surface layer of the jet, since the speed will not be uniform across the section. In the case of water individual determinations of γ ranged between 71 and 76 dynes/cm. and for mercury the variation was larger.

This of course would be a dynamic method, but there are obvious experimental difficulties in the way of making it a precise method.

(7) *Bond's flowing sheet method.* (*Proc. Phys. Soc.* **47**, 549, 1935.) Two accurately aligned vertical jets are opposed so that the impinging streams form a flowing sheet, which is horizontal at the centre but curving downward under gravity when farther from the jets. For a given rate of flow the maximum diameter attained by the sheet before breaking into drops is inversely proportional to γ.

Bond applied his formula to measurements on water, using jets of several sizes, and obtained values lying between 72·9 and 73·7 dynes/cm. for 20° C.

Puls (*Phil. Mag.* **22**, 970, 1936) applied the method to mercury, but individual readings varied widely and it is not possible to say how much of the variation is due to experimental uncertainty and how much to change in the quality of the mercury. In a subsequent paper Bond and Puls (*Phil. Mag.* **24**, 864, 1937) describe the use of this method as well as that of the oscillating jet to follow changes occurring in the surface tension of solutions during the first few thousandths of a second after formation.

(8) *Adhesion of rings and plates.* The method of finding γ from the force needed to lift a narrow glass slide or a horizontal ring of wire from a liquid surface has been widely used, and practical torsion balances for rapid determinations have been devised by du Nouy (*Jour. Gen. Physiol.* **1**, 521, 1919) and others. Errors by these methods may be considerable, due chiefly to uncertainty as to the weight of liquid supported by the plate or wire at the time of observation. Lenard's method (*Ann. Physik*, **74**, 381, 1924) of using a very thin wire tightly stretched across a frame has been found successful for students' use, the corrections being very small when the wire is clean and wetted by the liquid. The maximum pull as the thin horizontal wire comes through the surface is very definite and, without its ultimate refinements, the method is very simple and convenient for use in teaching.

(9) *Shape of pendent drop.* Many references occur throughout the literature to evaluations of γ based on the fundamental relation between curvature, surface tension and pressure difference on the sides of a film. In recent years a number of workers, following Andreas, Hauser and Tucker (*Jour. Phys. Chem.* **42**, 1001, 1938) have made determinations of surface and interfacial tensions from observations on pendent drops. The drop (Fig. 7) hangs from the

end of a tube having a diameter of from 1·5 to 2 mm. Photography gives an enlarged undistorted image of the drop and measurements are made of D, the maximum diameter of the drop, and also of the diameter d at a distance D above the lowest point on the drop. The ratio d/D is defined as the shape-factor, S, for the drop. The equation $\gamma = \rho g D^2/H$ gives the value of the surface tension, H being a function of S. Values of H for a range of values of S are given, having been determined from pendent drops of water. The authors consider the method should give results of very high precision

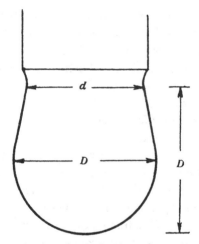

Fig. 7. Pendent drop: the shape-factor $S = d/D$

when the relation between γ and S has been determined more precisely. The method is a static one, independent of contact angle, and may be used for finding interfacial tensions also. It requires but a small quantity of the substance and is suitable for following slow changes in a surface. Distortion of the shape by the window or optical system may be detected by photographing an object of known dimensions at the position to be occupied by the drop.

Smith and Sorg (*Jour. Phys. Chem.* 45, 671, 1941) have used this method with a number of liquids, including water and benzene. For standard liquids their values are slightly higher than those in the tables, and for water almost one dyne/cm. higher. The significance of this is scarcely clear, since they use the values of H computed by the earlier workers from observations on pendent drops of water. Smith has further modified the technique in applying it to determine variation in surface tension with molecular weight and configuration in series of liquid hydrocarbons (*Jour. Phys. Chem.* **48**, 168, 1944).

3. Measurement of Small Differences in Surface Tension

(1) *Differential capillarimeter.* (*Jour. Am. Chem. Soc.* **59**, 187, 1937.) Jones and Ray describe a U-tube of fused silica, one arm being wide and accurately cylindrical, the other narrow (0·0136 cm. in radius). By having on the narrow arm a reference mark to which the tube is always filled, any error due to variation in cross-section of the capillary tube is eliminated. The tube is filled to the reference mark with the first liquid and weighed. The process is repeated with the second liquid and the difference in weights enables the difference in heights at which the liquids stood in the wide arm to be calculated with great accuracy. Differences of less than 0·1 dyne/cm. between γ for water and for dilute solutions of KCl were measured in this way.

(2) *Twin ring surface tensiometer.* A very sensitive method for measuring small differences in surface tension has been described by Dole and Swartout (*Jour. Am. Chem. Soc.* **62**, 3039, 1940). A frame in the form of a double ring is suspended from each arm of a sensitive balance. One frame dips into each of the two liquids to be compared and the dishes containing the liquids are then lowered till the rings are on the point of leaving the surface. The balance is used only to measure the difference in the forces necessary to pull the rings from the surfaces. This differential method eliminates most of the uncertainties associated with the determination of γ by the ring method. Using it, Dole and Swartout obtained substantial confirmation of the observations of Jones and Ray on dilute solutions.

4. Interfacial Tension

In dealing with the spreading of a liquid over the surface of another liquid or a solid the surface energy at the interface becomes as important as that at the free surface of a liquid, while in the case of emulsions stability obviously depends entirely on conditions existing at interfaces.

Interfacial tensions have been measured chiefly by the drop-weight method, though measurement of the large drop and of capillary rise has been used. The balance of du Nouy [2 (8)] may also be used to measure the force to pull the wire through an interface provided the ring is wetted by the lower liquid. Where it is desired to follow the variation of interfacial tension with time it is probable that the method of the large drop will be more widely used in future.

CHAPTER III

THE SURFACE OF LIQUID METALS: MERCURY

1. Metals in General

The values of surface energy for liquid metals are very high, ranging from 290 ergs/cm.^{2} for sodium to 1200 ergs/cm.^{2} or more for iron at 1300° C. and up to 1800 ergs/cm.^{2} for platinum. A large value for the free energy of a surface indicates a corresponding tendency for the adsorption to the surface of impurities that will lower the free energy. This adsorption may occur either from the vapour phase or from the liquid metal itself, the latter case being particularly important, since generally it has not been possible for experimenters to obtain specimens of metals having a purity comparable with that of distilled water.

As we have already mentioned [Ch. II, 2 (3)], the method of maximum bubble pressure seems most useful if a neutral gas is used for blowing the bubbles. Thus a new surface is formed in a neutral gas with each bubble and the results will probably differ but slightly from those for a surface in vacuum.

The method of the large drop [Ch. II, 2 (4)] may be used for metals either in gases or vacuum. [Attempts have been made to find the surface tension of a metal at its melting point by allowing a large drop to solidify and then measuring it. As most substances undergo a sudden change in volume on solidifying, this adaptation of the method is not of much use.]

For mercury almost all the methods described in the previous chapter have been used at various times.

Metals which are easily oxidized or which contain oxidizable impurities should be measured in a neutral atmosphere or else in high vacuum after special precautions to keep out oxygen.

At high temperatures metals will be less liable to contamination by volatile impurities than has been the case with mercury. Up to the present time, however, sufficient work has not been done with other metals to show whether the question of their surface tensions will be settled any more easily than that of mercury. Bircumshaw, who, using the method of bubble pressure, has done much work in this field, says that it is difficult to obtain concordant values for γ at low temperatures either for pure metals or for alloys.

References to the more recent papers on the surface tension of alloys are given by Bircumshaw (*Phil. Mag.* **17**, 181, 1934). In this paper he records that in lead-tin alloys the lead produces a lowering of surface tension. A small amount of lead added to tin produces a considerable reduction in γ, whereas the addition of tin to lead produces but little rise. This is an illustration of the general principle that in any liquid phase the molecular species yielding the lowest surface energy will be selectively adsorbed at the surface.

2. Mercury

(1) *The remarkable discordance in observations.* It is not difficult to obtain mercury in which the percentage of impurity is extremely small, and this metal has naturally been a favourite subject of investigation. Hence one might expect mercury to be a standard substance for relative determinations on metals and that its surface tension should be known to about the same degree of accuracy as that of water or benzene. Far otherwise are the facts of the case. Comparatively recent papers, describing measurements which have apparently been made with the highest precision on mercury in vacuum, yield values ranging from 430 to 515 dynes/cm., though any individual worker generally obtains readings which indicate an accuracy of about ¼ of 1 %.

A critical analysis of the available data and the methods used, however, makes it reasonably certain that γ for mercury is not less than 480 dynes/cm. and probably it is higher than this figure.

A mass of work has been done on the apparently large variation of the surface tension of mercury with time of exposure to various gases, and a wealth of ingenuity expended in explaining the observations, which range (in the case of exposure to air) from a fall of 40 or 50 dynes/cm. in a few minutes down to as little as 5 dynes/cm. in 24 hours.

For the interfacial tension against water and other liquids the values likewise are spread over a wide range but, apart from electro-capillary phenomena, interfacial tension is chiefly of interest in the discussion of spreading. The interfacial tension of mercury against water will be considered in this connection in Ch.v.

A history of observations extending over more than a century on the surface properties of mercury would be of interest from several points of view, but here we can give only an indication of the growth of our present ideas on the subject and of how far some of these are from final confirmation.

(2) *Surface tension in vacuum and in gases.* For a long while there was a certain amount of agreement that for mercury in vacuum the value of γ was very close to 436 dynes/cm. and this figure is still to be found in some texts and sets of tables.

For a surface formed in gas at atmospheric pressure values of over 550 dynes/cm. are on record, γ falling rapidly at first and then more slowly to the value for vacuum or lower. The rate at which γ fell in value depended on the nature and pressure of the gas. The higher value for surface energy in the presence of a gas was accepted, but it presents an obvious difficulty, for how can the gas act except by being adsorbed on the surface, and does not adsorption occur just because the process can lower (not raise) the free energy? Various explanations have been put forward, but the most reasonable view seems to be that put forward by Iredale, viz. that the low value in vacuum indicated a contaminated surface and that the presence of the gas hindered the contamination reaching the surface so quickly as it did when the space above the surface was evacuated.

[Other suggestions dealt with the possibility of the molecules of gas retarding either the orientation of the molecules of mercury to positions of least energy or the adsorption to the surface of those molecules (from the bulk of the liquid) which would give minimum surface energy.]

Much of the confusing data in the literature has come from workers who employed the method of the large drop, but this method seems the natural one for following changes in the surface occurring over a period of time. Recent determinations by this method, after the most careful out-gassing of the apparatus, have yielded values for vacuum as high as or higher than the initial values in the presence of gas.

In spite of many attempts the writer has never succeeded in observing the low value in vacuum even when the apparatus purposely had been made dirty, and the actual cause of the low values, which some have undoubtedly observed, remains unsettled.

In the table are given some of the more recent determinations for mercury. These appear to offer a basis for setting at least a lower limit to the possible value of γ for mercury, though it must be said that other papers indicating a far lower value are to be found in the literature (cf. Popesco, *Ann. de Physique*, **3**, 402, 1925 and Kernaghan, *Phys. Rev.* **37**, 990, 1931).

Author and Method	γ (Vacuum)	γ (Gases)	$d\gamma/dT$ and range
(1) Harkins and Ewing (Drop-weight)	476	464 (air)	0·22 (0–60° C.)
(2) Bircumshaw (Bubble Pressure)	—	480 (H_2 at 20° C.)*	0·2 (0–20° C.)
(3) Cook (Large Drop)	Up to 515 at 31° C.	Slightly lower than in vacuum	—
(4) Burdon (Large Drop)	488 at 25° C.	Same as in vacuum	0·23 (20–230° C.)
(5) Bradley (Large Drop)	500·3 at 16·5° C.	Very slow fall in H_2	—
(6) Bosworth (Bubble Pressure)	—	484 (air at 20° C.)	0·21 (20–260° C.)
(7) Kemball (Large Drop)	484·2 at 25° C.	—	0·2 (25–75° C.)

(1) *Jour. Am. Chem. Soc.* **42**, 2539, 1920.
(2) *Phil. Mag.* **12**, 593, 1931.
(3) *Phys. Rev.* **34**, 513, 1929.
(4) *Trans. Far. Soc.* **28**, 866, 1932.
(5) *Jour. Phys. Chem.* **38**, 231, 1934.
(6) *Trans. Far. Soc.* **34**, 1501, 1938.
(7) *Trans. Far. Soc.* **42**, 526, 1946.

* Bircumshaw records a value lower by several dynes when the hydrogen bubble was held on the tube for 30 min. before detachment.

The recent careful work of Kemball makes it seem fairly certain that γ for mercury at 20° C. lies within one per cent of the value 485 dynes/cm. In view of the great liability to contamination of the surface one feels that higher values are more likely to be correct than lower ones, but Kemball discusses very fully the possible sources of error and his claim to an absolute accuracy of $\pm 1·5$ dynes/cm. seems reasonable. As pointed out by the writer (*Nature*, **128**, 456, 1931) very high values obtained from the sessile drop might contain an error due to viewing a reflection in the upper surface of the drop through a telescope having its optical axis not truly horizontal.

(3) *Capillary depression and meniscus volume.* Various tables are given in the literature for the corrections necessary in working with manometers to allow for the effect of the meniscus. In view of the

wide variations in γ produced by traces of contamination it is only to be expected that these tables are of limited value. Kistemaker (*Physica*, **11**, 270, 1945) reviews work on this subject and suggests a method of checking the amount of depression by taking readings on tubes having different diameters. By this method he considers the depression for a tube of 1·8 cm. diameter should be determined to within ± 0·003 mm.

(4) *Temperature coefficient of* γ. Not many observations are available on the variation of surface tension of mercury with temperature. The figures given in the table above for $d\gamma/dT$ are in much better agreement than those for the value of γ. It seems certain that the temperature coefficient must increase at higher temperature, or the critical temperature for mercury would be above 2000° C. Hogness (*Jour. Am. Chem. Soc.* **43**, 1621, 1921) used a method akin to the bubble-pressure method by measuring the pressure necessary to form drops of various molten metals on the tip of a capillary tube of quartz. For mercury he found no difference between the values for the surface tension in vacuum, in hydrogen, and in dry air. Hogness's readings, which were taken up to 350° C. in the case of mercury, give values for $d\gamma/dT$ which increase continually with temperature, being lower than those given above for temperatures up to 200° C., but considerably above them for higher temperatures.

On the other hand, Kemball's work suggests that γ for mercury falls by a larger amount between 25° C. and 50° C. than it does between 50° C. and 75° C. This is probably not significant, as the temperature range is small and the variation in γ of almost the same order as the possible error.

A satisfactory determination of the temperature coefficient through a large range of temperature would be of interest. An earlier suggestion by the author to use the method of capillary depression in a U-tube having one wide and one narrow arm sounds less reasonable in view of a paper by Bate (*Phil. Mag.* **28**, 252, 1939). In this paper Bate describes the measurement of γ for mercury in a glass tube with the surface of which the mercury made a contact angle of 90°. In an apparatus made of such glass there would of course be no capillary depression. Bate's value for γ was 490 dynes/cm. There are other references in the literature to the variability of contact angle of mercury with glass, depending on the treatment of the surfaces.

(5) *Gases and vapours on the surface of mercury.*

(a) *Effects on surface tension.* Using the large drop, Popesco (*Ann. de Physique*, **3**, 402, 1925) obtained a constant low value for the surface tension in vacuum, and in gases a high initial value, falling rapidly at first and then more slowly to the value obtained in vacuum or even lower. Earlier work was in general agreement with this (e.g. Stöckle, *Wied. Ann.* **66**, 499, 1898).

Some years ago the writer made observations on a large drop of mercury in a glass apparatus and obtained a rate of fall of surface tension in filtered air that was in fairly close agreement with the curves given by Popesco. When the drop was formed in unfiltered air it was found that, during the first five minutes, γ fell about three times as quickly as in the filtered air.

Somewhat more recently (*Trans. Far. Soc.* **28**, 866, 1932) a drop formed in a quartz apparatus that had been degassed at red heat was found to fall by only 4 or 5 dynes in 24 hours' exposure to dry air. The rate of fall of γ increased thirty-fold when the drop was formed in air that had been bubbled through water. When the apparatus was wet a surface formed in moist air from the laboratory fell by 50 dynes in half an hour. Experiments with this apparatus made it clear that water vapour could not account for the low values that some workers had found for mercury in moderate vacuum, since in an atmosphere of water vapour the *initial* value was found to be nearly the same as for vacuum or dry air.

It seems fairly certain that the *rapid* fall in γ observed by various workers, who formed large drops of mercury in the presence of different gases, has been due to some impurity rather than to the gas itself. Of course very slight variations in the purity of the mercury itself might be responsible for large variations in surface tension, e.g. where the impurity can react with oxygen in the gas, forming a layer of oxide on the surface.

Cook (*loc. cit.*), using the apparatus with which he determined the surface tension of mercury, observed a slow fall in γ for a large drop of mercury in the presence of hydrogen at low pressure (0·005 mm.), the reduction in γ being accompanied by a lowering of the pressure of hydrogen. Bumping the apparatus so as to break the surface of the mercury caused a partial recovery both in the pressure of the hydrogen and in the value of the surface tension, though the latter never rose above 455 dynes/cm. (as against a value of about 500 dynes/cm. for the freshly-formed drop). The observations on adsorption of the hydrogen would be qualitative only, owing to uncertainty as to the amount of hydrogen which had been adsorbed by the glass, and not by the mercury.

Bosworth reports some work carried out by the drop-weight method (*Trans. Far. Soc.* **35**, 1353, 1939). Using drops of mercury formed in air containing varying percentages of CO_2, he finds a progressive fall in γ till a concentration of 3 % CO_2 is reached. Beyond this γ rises until, in pure carbon dioxide, the value is within 1 % of that for mercury in pure air.

Kemball and Rideal (*Proc. Roy. Soc.* A, **187**, 53, 1946) have investigated the reversible adsorption of benzene and other non-polar vapours on mercury. The observed reduction in γ depends on the pressure of the vapour and indicates that the adsorbed molecules behave as an imperfect two-dimensional gas (Ch. VI, 4).

(*b*) *Degree of persistence of adsorbed gases.* Oliphant (*Phil. Mag.* **6**, 422, 1928) and Bosworth (*Trans. Far. Soc.* **28**, 896, 1932) showed that a shower of mercury drops falling through a mixture of gases (e.g. 2 or 3 % CO_2 in argon or hydrogen) at atmospheric pressure would selectively adsorb practically a monomolecular layer of CO_2 within a fraction of a second of the formation of the drops.

The author (*Proc. Phys. Soc.* **47**, 460, 1935) formed surfaces of mercury in gases at various pressures, rapidly exhausted the vessel to 10^{-4} mm. and then measured the gas evolved when the surface was caused to collapse by allowing the mercury to flow out. Here the gas corresponded approximately to a monolayer over the mercury. Bumping the apparatus so as to cause an oscillation of the surface did not cause the gas to be freed from the surface. Check experiments showed that adsorption on the apparatus was not playing any part in these results, but it was found that if the mercury surface was formed in this moderate vacuum (10^{-4} mm.) and the gas then admitted there was practically no evidence of adsorption occurring. This behaviour is in keeping with observations fairly commonly recorded a few years ago of the difference between an expanding or nascent surface of mercury and one at rest. For instance, a drop of liquid that will not spread over a surface of mercury at rest may be carried out over the surface as it expands. Such observations have encouraged the search for differences between the static and dynamic surface tension in mercury. It seems likely however that the differences are due chiefly to rapidly acquired contamination of the surface.

(c) *Conclusions.* It must be admitted that very little can be said with any high degree of probability concerning the action between gases and the surface of mercury. It is fairly established that in the presence of many gases the surface of mercury adsorbs a monolayer of gas practically during the process of formation of the surface and that more active gases can displace less active ones from the surface (Bosworth, *loc. cit.*). No measurable fall of surface tension appears to occur in the short time required for the adsorption of this gas layer, a fact which agrees with the observation that the gas is easily displaced from the surface. Observations recorded over many years, of the fall of surface tension with time of exposure to gases, differ so widely that nothing definite can be said either as to amount or cause of the fall in γ.

One readily seconds the appeal of N. K. Adam that future workers on the surface phenomena of mercury should take the most elaborate precautions to secure purity of mercury, degassing, etc., and should give an account of their methods that will be sufficient to enable their observations to be compared with those of other workers.

The hope that surface phenomena on mercury will be elucidated by means of electron diffraction does not receive very much support from the attempts that have so far been made. The presence of mercury vapour tends to destroy the definition of the beam of electrons and of course the beam may be expected simply to remove a loosely adsorbed layer rather than reveal its presence. (Cf. Bailey, Fordham and Tyson, *Proc. Phys. Soc.* **50**, 63, 1938.) Observations on surface layers have been made by applying Drude's theory of the reflection of polarized light, but the results so far have not been very conclusive in the case of mercury.

3. Other Liquid Metals: Gallium

From the position with regard to mercury it is clear that precise and reliable determinations of the surface properties of molten metals are improbable at present. There seems no reason to expect that they will prove less liable than mercury to have their surface energies affected by adsorption. As with mercury there may be adsorption of impurities in the metal, or of gases, as well as reactions between gases and impurities. It is perhaps as well that other molten metals are not so convenient as mercury for the 'trying out' of methods of measuring surface tension. This use of mercury for trial purposes has certainly been responsible for papers which add to the confusion when one tries to arrive at the most probable value of γ for mercury.

Davis, Mack, and Bartell (*Jour. Phys. Chem.* **45**, 846, 1941) have applied the method of the pendent drop to find the surface tension of gallium, a metal which melts at 30° C. Near its melting point they got the value 735 ± 29 dynes/cm. for the surface tension of gallium in a neutral atmosphere; the admission of air reduced the value to approximately 300 dynes/cm. For an atmosphere of CO_2 Davis, Mack and Bartell's value is more than twice as great as that given in International Critical Tables, a remarkable illustration of the uncertainty of published data in this region.

A determination of γ for mercury, using exactly the same set up as for the measurements on gallium would have been of interest and value since the uncertainty in the value of γ for mercury is not now very great.

CHAPTER IV

SPREADING: GENERAL CONDITIONS

1. No Simple and Complete Theory of Spreading

There are probably few branches of scientific investigation where the tendency to generalize from too few observations or apply too widely an explanation that fits a particular group of phenomena has been more evident than in the study of the spreading of one liquid over the surface of another.

Until recently the theory persisted that a drop of one pure liquid would always spread completely over the surface of another pure liquid on which it was placed. When an appeal to experiment yielded no evidence of spreading, it was assumed either (1) that the liquid surface was contaminated or (2) that a layer too thin to be detected had in fact spread over the surface and that the remainder of the drop was in equilibrium with this film. However, Hardy (*Proc. Roy. Soc.* A, **86**, 610, 1912) reported that pure heavy paraffins did not spread on water, and he was apparently the first to suggest that chemical affinities between the liquids might be important as a cause of spreading.

Langmuir (*Jour. Am. Chem. Soc.* **39**, 1848, 1917 and *Trans. Far. Soc.* **17**, 673, 1922) gave an account of the chemical theory of spreading, particularly in connection with his work on monomolecular films (Ch. VI). Following these papers again came the tendency to make the theory all-embracing; spreading was held to be due to forces exerted only between one molecule and an immediate neighbour, and the result of spreading should always be a single layer of molecules spread over the lower liquid, with the excess of the upper liquid collected into drops on the monomolecular film; completely saturated substances of course should not spread. Actually the purest paraffins spread readily on mercury and neither here nor in the spreading of water over mercury has the existence of a monomolecular film been established.

Edser (*Fourth Report on Colloid Chemistry*, B.A.A.S.) championed what might be called the *physical* theory of spreading. According to Edser, a molecule is to be regarded as surrounded by a field of force which decreases so rapidly with distance that in effect 94% of the force is exerted on contiguous molecules and only the balance on molecules beyond the first layer. This balance of force would still be large compared with gravitational forces and the spreading is thus due to a squeezing out of the liquid due to forces extending beyond the first layer of molecules. Similar ideas as to the cause of spreading had been put forward over a century ago, but the idea of a simple field of force does not take us far, since we know that in many cases spreading depends on the properties of parts of the molecule rather than on its properties as a whole.

For the spreading of a liquid over a solid, Hardy propounded the theory that spreading occurred through the condensation of vapour from the liquid over the solid and the subsequent spreading of the bulk of the liquid over this film of condensed vapour (*Proc. Roy. Soc.* A, **100**, 550, 1922). Attempts were made to extend this 'distillation' theory to the case of spreading over liquids, but without any good reason.

Probably all the various factors which have been suggested from time to time as the cause of spreading play some part in the phenomenon, different factors predominating in different cases.

2. Energy Conditions

The necessary condition for spreading to be possible is simply that the process shall involve a reduction in free energy.

Consider a drop of liquid A resting on a liquid B (Fig. 8). If A spreads to cover an additional square centimetre of B, then this has involved an increase of 1 sq.cm. in the area of the upper surface of A and of the interface, and the disappearance of 1 sq.cm. of the surface B. Hence for spreading to be possible $\gamma_A - \gamma_B + \gamma_{AB}$ must be negative. Harkins's spreading coefficient, which must be positive for spreading to occur, is

$$S_{AB} = \gamma_B - \gamma_A - \gamma_{AB}. \qquad (1)$$

If this spreading coefficient has a positive value, then

$$\gamma_A - \gamma_B + \gamma_{AB} < 0,$$

and still more $\qquad \gamma_A - \gamma_B - \gamma_{AB} < 0;$

hence if A can spread on B, then B cannot spread on A.

The work done in separating two liquids which are in contact over an area of 1 sq.cm. is called the work of adhesion of the two liquids. Since the separation involves the disappearance of 1 sq.cm. of interface and the appearance of a sq.cm. of free surface of each liquid, the increase in surface energy involved is clearly

$$\gamma_A + \gamma_B - \gamma_{AB}.$$

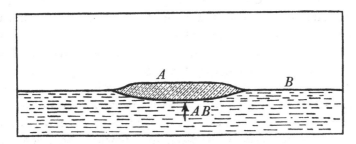

Fig. 8

In the same way the work of cohesion for liquid A would be expressed as $2\gamma_A$, since breaking a column of 1 sq.cm. cross-section involves merely the creation of 2 sq.cm. of new surface. Expressed in terms of adhesion and cohesion, we might expect A to spread on B if the adhesion of A for B is greater than the cohesion of A, i.e. if

$$\gamma_A + \gamma_B - \gamma_{AB} > 2\gamma_A$$

or
$$\gamma_B - \gamma_A - \gamma_{AB} > 0. \qquad (2)$$

But (2) is simply the condition that the spreading coefficient (1) shall be positive.

Experiment shows that in most cases where A spreads as a thick film (called 'duplex' by Harkins), the film later contracts to form lenses in equilibrium with a monolayer on the liquid B. It is probable that no other condition is thermodynamically stable (cf. Harkins, *Jour. Chem. Phys.* **9**, 552, 1941), but films of oil have been observed to persist on water for long periods without visible contraction (Ch. VI, 12). When the contraction of the thick film begins it of course does so because the contraction produces a further reduction in the free energy of the system. The spreading coefficient has now become negative, usually because of a reduction of γ_B with time. This reduction may be due to condensation of vapour from A on to B, or, where A is a solution of an active substance, to molecules escaping from A to form a monolayer of low surface energy on B, as observed by Mercer and others (Plate II).

Gravitational energy must also play some part in causing spreading, but in most cases this factor is almost negligible. Langmuir (*Jour. Chem. Phys.* **1**, 756, 1933) has considered the equilibrium of large lenses of oil on the surface of water and has taken account both of the effect of gravitational forces and of the curvature of the surface of the lens of oil in opposing spreading (cf. the theory of the large drop, Ch. II).

Though the *necessary* condition for spreading over liquid or solid is simply that the process shall involve a decrease in free energy, this is as general as the statement that water flows downhill—a dam may prevent the water from flowing, and a potential barrier may prevent the liquid from spreading, even though the free energy would be reduced if the liquid did succeed in spreading. (See Spreading on Solids, § 4.)

When spreading is possible, it is the mechanism of the process rather than the magnitude of the spreading coefficient (decrease in free energy) that determines the rate of spreading. Harkins has pointed out that spreading may be extremely rapid in spite of a small coefficient, and in other cases very sluggish in spite of much larger energy differences.

In cases of non-spreading liquids the addition of a minute quantity of some 'active' substance to A may produce spreading, because selective adsorption of the active substance at the interface reduces the value of γ_{AB}. Thus a trace of fatty acid dissolved in a heavy paraffin will cause that liquid to spread on water (Ch. VI), and the effect is shown to a remarkable degree in the spreading of water on mercury (Ch. V). In the latter case the spreading is completely determined by adsorption of ions at the interface.

3. Mechanism of Spreading on Liquids

In keeping with the idea of an elastic film over the surface of a liquid is the popular view that an oil spreads on water because it lowers the surface tension and the stronger surrounding film on the water then contracts, pulling out the oil. The propulsion of celluloid toys over water by a scrap of camphor sealed to them just at the waterline, and other such phenomena, lend an air of verisimilitude to this idea of a contractile skin. But if the water is dusted with talc and a drop of a spreading oil is placed on it, the talc is not moved until the advancing edge of the oil actually reaches it. The oil sweeps the talc ahead of it; there is no evidence of a skin contracting and carrying the dust with it.

When the molecules of the spreading substance (e.g. oil) have a definite affinity for the lower liquid (e.g. water), the mechanism is apparently somewhat as follows (cf. Langmuir, *loc. cit.* and Adam, *Physics and Chemistry of Surfaces*, p. 212). Molecules of oil become attached to molecules of water. Then molecular agitation in the water allows more and more molecules of oil to reach the water, and a layer of molecules of oil thus diffuses rapidly over the surface. Cohesion and viscosity in the oil cause these adhering molecules to drag a layer of oil with them, thus spreading the drop as a whole. In many cases only the single adhering layer of molecules is stable, and the excess oil collects again into one or more drops or lenses. Globules of fat seen floating on water are usually such lenses in equilibrium with a monomolecular layer of fat covering the rest of the surface.

The view that the only stable spreading is one which results in a monomolecular layer in equilibrium with drops is perhaps too comprehensive. Pure paraffins on mercury have not been observed to behave in this way and it is at least possible that in other cases multimolecular films may be formed owing to a kind of induced polarity due to the field of the lower liquid (cf. hanging a number of iron nails end to end from a magnet).

The spreading mechanism described is quite satisfactory for molecules possessing a definite active end, e.g. fatty acids of the type $CH_3(CH_2)_n COOH$. The lowest members of the series dissolve in water, but when the number of CH_2 groups increases, the tendency of the head (COOH) to dissolve no longer suffices to pull the whole molecule into the water. The heads adhere firmly to the water and when the molecules in this first layer are closely packed the upper surface would consist solely of the saturated CH_2 groups. Over such a layer a drop of benzene cannot spread, even though benzene will spread over and dissolve the solid paraffins.

An essential in the mechanism of spreading over liquids is the mobility of the underlying liquid. It is the molecular agitation and diffusion in the water that spreads the oil. Spreading does in fact occur on a liquid whenever the energy conditions are favourable.

An active·solid floating on a liquid would also spread by this mechanism, though of course only to a monomolecular layer, since the adhering molecules could not pull out other molecules of the solid along with them.

This mechanism also accounts for the effect of a trace of stearic acid in a non-spreading paraffin. Molecular agitation and diffusion in the paraffin allow the molecules of stearic acid to reach the water surface where they adhere. When all the active molecules have reached the surface, then the spreading ceases and the paraffin usually contracts to form lenses on the monolayer (Ch. VI).

4. Spreading on Solids

(1) *Energy condition: angle of contact.* As we have already seen (§2), spontaneous spreading over a surface can occur only when it results in a reduction in the free energy of the system, but the free energy of the surface of a solid cannot be measured directly. Assuming that the surface of a solid is perfectly uniform (a condition which never actually holds), then if a drop of pure liquid is placed on a horizontal surface it may spread out indefinitely, or it may spread only to a limited area and come to rest so that the film at the edge of the drop everywhere meets the surface at a definite angle called the angle of contact. It is fairly obvious that the more nearly this angle approaches zero the more nearly will the liquid spread completely over the surface.

If the drop of liquid A covers an extra sq.cm. of B, the free energy lost in the process is

$$\gamma_B - \gamma_A - \gamma_{AB}.$$

When the drop is in equilibrium, making an angle of contact θ, we have
$$\gamma_B = \gamma_{AB} + \gamma_A \cos \theta, \qquad (1)$$

but γ_A and $\cos \theta$ are the only measureable quantities (Ch. VII).

However, let us revert to the idea that spreading occurs when the adhesion of liquid for solid, $\gamma_A + \gamma_B - \gamma_{AB}$, exceeds the cohesion of the liquid, $2\gamma_A$.

Using relation (1) to substitute for the quantity $\gamma_B - \gamma_{AB}$, we have for the work of adhesion of A to B the expression $\gamma_A (1 + \cos\theta)$, which contains only directly measurable quantities.

Obviously $2\gamma_A > \gamma_A (1 + \cos\theta)$ so long as θ is finite. Consequently spreading entails the condition that $\theta = 0$ or that the adhesion is as great as the cohesion. Of course the work of adhesion between liquid and solid may exceed that of cohesion of the liquid, but this condition cannot be expressed in terms of contact angle. Harkins (*Jour. Chem. Phys.* 10, 342, 1942) stresses this point in a comprehensive treatment of the work of adhesion between liquid and solid, and shows that the expression $\gamma_A (1 + \cos\theta)$ should be regarded as the decrease in free energy when A spreads to cover an additional sq.cm. of the surface of B which is already in equilibrium with the saturated vapour of A. This generally is much less than the loss of free energy of the solid surface due to adsorption of a layer from the vapour. For some crystalline solids, such as graphite, where the ratio of surface to volume can be made sufficiently great, direct calorimetric measurements will give evidence of the work of adhesion of liquid to solid.

For complete spreading or wetting of the solid, the contact angle must be zero. In a capillary tube the liquid will be raised or depressed with respect to the level on the free surface of the liquid according as the angle of contact between the liquid and the wall of the tube is less or greater than 90°.

The quantity $\gamma_A \cos\theta$ is often called the adhesion tension. It is this quantity that determines how strongly the liquid will be drawn into a capillary in the solid.

(2) *Mechanical conditions for spreading on solids.*

(a) *Effect of immobility of surface of a solid.* For the process of spreading over a liquid we saw that fluidity of the lower liquid was of prime importance, since it was diffusion in this that caused the adhering molecules of the upper liquid to be carried outward. This factor cannot operate in the case of a solid, and its absence changes the mechanism of spreading in such a way that, generally speaking, 'active' substances do not spread readily on a solid and may in fact be used to retard the spreading of the less active components of a liquid. For instance, the COOH groups in the molecules of fatty acids become fixed at a metal surface, thus yielding an immobilized layer of molecules having their saturated ends outward. Over this layer spreading does not readily occur. In keeping with this, Woog (*Compt. Rend.* **181**, 772, 1925) found that over either water or mercury the unsaturated (active) oils spread most rapidly, whereas on the solid metals the saturated oils showed more spreading. The persistent creeping of transformer oils over metals is well known, and Bulkley and Snyder (*Jour. Am. Chem. Soc.* **55**, 194, 1933) report a spreading of pure paraffins over clean steel even though the resulting reduction in free energy is much less than would result from covering the surface with an active substance that would not spontaneously spread on it.

(*b*) *The 'vapour-pressure' theory of spreading.* Hardy (*Proc. Roy. Soc.* A, **100**, 573, 1922) found that if a drop of a lubricant having an appreciable vapour pressure were placed on a surface of glass its lubricating effect could be detected by a reduction of friction at some distance from the visible drop. Such experiments gave rise to what has been called the distillation theory of spreading, viz. that a primary film is formed over the solid by condensation from the vapour phase and that the thicker film subsequently spreads over this primary film. Though Hardy's experiments establish the reality of this process, the suggestion that this might be the universal mechanism of spreading on both solids and liquids is hardly warranted. It is at least doubtful whether the creeping of paraffins on solids has this explanation, and their very rapid spreading on mercury certainly has not. Mercury is found to spread slowly over a copper surface immersed in oil when vapour cannot play any part, and Nietz (*Jour. Phys. Chem.* **32**, 265, 1928) found that the spreading of an active solid on water was stopped by covering the vessel so that the space became saturated with vapour of the solid. A drop of distilled water spreading over mercury is stopped by saturating the air with water vapour. In this case the vapour condensing on the surface reduces the surface tension sufficiently to make the spreading coefficient zero or negative. On liquids generally, condensation from the vapour phase appears to retard spreading, as in fact would be expected from the reduction of γ_B in the spreading coefficient.

(3) *Other factors in spreading on solids.*

(a) *Surface migration.* Since the work of Volmer and Estermann (*Zeit. f. Physik*, **7**, 13, 1921) there has been a growing recognition that the surface of a solid and the molecules adsorbed thereto constitute a much less static system than had previously been pictured. Observing mercury crystallizing from its vapour, they found that the growth of the crystals could be explained only if all the molecules which struck the crystal condensed and then travelled along the face of the crystal to settle down on the growing edge. It is now well known that at higher temperatures molecules may diffuse over the surface of solids. Bosworth, for instance (*Proc. Roy. Soc.* A, **154**, 112, 1936), has followed the diffusion of alkaline metals over the surface of tungsten by means of the photo-electric effect, which could be set up only on parts to which the alkaline metal had spread.

Apparently molecules in the surface of a solid, or attached thereto, may be raised in potential energy to a level where they can slide over the surface though they have not the energy necessary to escape from it. As a solid approaches melting point the molecules occupying points of high potential (edges and corners) are able to travel to areas of lower potential.

This migration of molecules over solid surfaces must play some part in assisting the slow spreading of a layer of molecules, but in the phenomena with which we are concerned here the effect would be insignificant.

(b) *Gravitational forces.* Where we have spreading only to a disc of finite size gravitational energy must be taken into account as a factor in the final equilibrium, but in the case of complete spreading this factor is negligible.

(c) *Long-range molecular forces.* Lubrication phenomena (Ch. VII) give some evidence of molecules exerting forces over greater distances than those separating them from their immediate neighbours, so some squeezing action may contribute to the spreading of liquids over solids. Thus, if forces exist between molecules in the superficial layer of the solid and molecules in the second, third, and more distant layers in the liquid, then these forces will assist spreading by squeezing out the intervening layers of fluids (cf. Ch. IV, 1).

CHAPTER V

SPREADING ON THE SURFACE OF MERCURY

1. Phenomena of Spreading

The mere observation of the phenomena of spreading is a fascinating occupation by reason of the diversity and interest of the effects that may be produced. The discovery of new phenomena in this region, however, is much easier than finding even a qualitative explanation for them. The high surface tension of mercury makes it possible for a liquid to possess a very big spreading coefficient against this substance. Of those given by Harkins (*Jour. Am. Chem. Soc.* **42**, 2543, 1920) water has the lowest coefficient, but even here the value given is 32 ergs/cm.[2]. However, interfacial tension is the deciding factor in the spreading coefficient, and as we shall show later, there is some doubt as to what significance can be attached to measurement of the interfacial tension of water and mercury.

High-grade conductivity water spreads but slowly over the surface of mercury and does not spread at all unless the surface is remarkably free from contamination (which would lower γ and so reduce the spreading coefficient). On clean mercury in air the spreading of water is greatly accelerated by the addition of as little as one part of inorganic acid or neutral salt to a million parts of water, while stronger solutions flash across the surface and frequently present remarkable subsequent expansions and contractions (*Proc. Phys. Soc.* **28**, 148, 1926).

On the other hand, the presence of traces of alkalis (such as NaOH, NH_4OH, $Ca(OH)_2$, etc.) in the water will prevent any spreading. Aqueous solutions of such alkalis are the only liquids which the writer has observed not to spread on clean mercury. A drop of solution of ammonia placed on mercury exposed to laboratory air remains as a non-spreading lens for a period which may be of some minutes' duration. When loss of NH_3 and absorption of CO_2 from the air has sufficiently changed the composition of the drop the edges flatten and it spreads fairly rapidly over the surface.

SPREADING ON THE SURFACE OF MERCURY

2. Spreading of Water and Aqueous Solutions

(1) *Spreading to a limited area.* Whereas a drop of purest water takes some minutes to spread to a disc of 2 or 3 in. diameter, a drop of solution of an organic acid of a concentration below one thousandth normal will spread to cover such an area in less than 1 sec. In this region was first observed the abrupt fall in the rate of spreading when the drop had covered an area very close to 1 sq.cm. for each 10^{14} molecules of monobasic acid present (*Trans. Far. Soc.* **23**, 205, 1927). The drop of dilute solution appears to spread

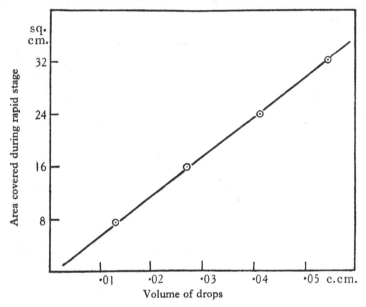

Fig. 9. Areas covered by drops of solution of constant concentration (9×10^{-5} normal HCl)

rapidly to a definite area and stop, though measurement shows that there is a subsequent spreading at about the same speed as that for distilled water. Curve *A* (Fig. 11), in which diameter of drop is plotted against time from Fuller's photographs of the spreading drop (*Trans. Far. Soc.* **33**, 1528, 1937), shows how abrupt is the change from rapid to slow spreading when the drop has reached the limiting area determined by its content of acid.

Using a constant concentration of acid (about 10^{-4} N), the area covered is strictly proportional to the mass of the drop (Fig. 9).

When drops of equal mass but varying concentrations were used, the area covered was very nearly proportional to the concentration (Fig. 10). Actually the readings show that the area covered increases slightly faster than the concentration, possibly a mechanical effect due to the greater speed of spreading of the more concentrated drops.

Aqueous solutions of any of the first five acids of the series $CH_3(CH_2)_nCOOH$ (i.e. from formic to valeric) show this same property of spreading rapidly to cover 1 sq.cm. for each 10^{14} molecules of acid in the drop. In fact at a dilution of 10^{-4} N they behave in a manner exactly similar to HCl or HNO_3 on a clean surface of mercury.

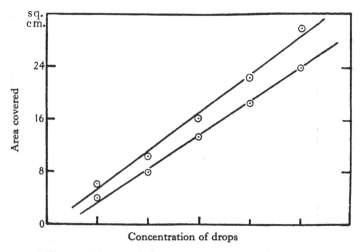

Fig. 10. Areas covered by drops of varying concentration

Sulphuric or oxalic acid covers twice as great an area per molecule. Thus, the area covered is determined by the number of H-ions obtainable from the acid, each H-ion apparently causing the water to cover rather more than 10 atoms of mercury in the surface layer. (1 sq.cm. of surface of mercury contains rather more than 10^{15} atoms.) The suggestion has been made that the definite area covered is due to a monolayer of ions with their groups of water molecules, but no way has been devised to test this.

Similarly stearic acid dissolved in paraffin may cause it to spread over water to an area several times as great as that required for a monolayer of stearic acid at the interface (Ch. VI, 6).

(2) *The speed and mechanism of spreading.* No satisfactory idea has been suggested for the mechanism by which practically every molecule of acid in the drop of solution can reach the interface during the rapid stage of the spreading. The speed of the process, coupled with the fact that only a few parts per million of the active spreading agent are present, suggests that here at any rate the forces are exerted through distances considerably greater than that between neighbouring molecules. The forces concerned are probably between ions rather than between neutral molecules. Observation shows that at the end of the rapid stage of spreading a good deal of the water in the drop is collected into a ridge round the edge of the area covered (Plate I). Some kind of squeezing action across the interface plus a resistance to flow at the periphery affords a rather plausible explanation of the water being forced into this peripheral ridge while the acid molecules reach the interface.

Efforts were made with the aid of a tiny pipette to collect water from the surface after the rapid stage had ended, in order to test if the acid remained on the mercury, but these attempts were inconclusive.

If we regard the spreading as depending on the ions or active molecules reaching the interface, then as long as the concentration of active molecules in the drop has not fallen appreciably, and so long as the concentration is uniform throughout, we should expect the number of new molecules reaching the interface in unit time to be proportional to the area, A, of the interface. This would give for the law of spreading

$$\frac{dA}{dt} \propto A, \tag{1}$$

a law which holds for the early stages of the spreading of paraffin oil containing traces of fatty acid over the surface of water (Ch. VI).

Such a simple relation is unlikely to hold for the spreading of a drop of aqueous solution to a limited area (Fig. 11 A), since presumably the concentration of active molecules falls very nearly to zero during the rapid stage. Actually both here and with a more concentrated drop, where concentration would not decrease appreciably during the spreading, the camera shows that the diameter increases at a nearly constant rate (Fig. 11 *A* and *B*). Thus for the actual rate of spreading we have

$$\frac{dr}{dt} = \text{constant},$$

and hence

$$\frac{dA}{dt} \propto \sqrt{A},$$

which does not fit any simple description in terms of diffusion and concentration.

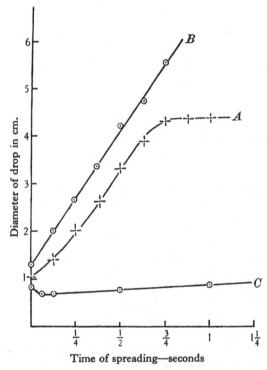

Fig. 11. Rate of spreading over mercury. *A*, Very dilute acid, limited spreading; *B*, more concentrated acid; *C*, distilled water

For the slow spreading of pure distilled water (Fig. 12*A*) the simple relation (1) might be expected to hold since the concentration of dissociated ions probably remains constant, any loss to the interface being replaced by further dissociation. The graph shows that (1) does not hold, but that the growth of *r* here too is linear.

PLATE I   *facing page* 46

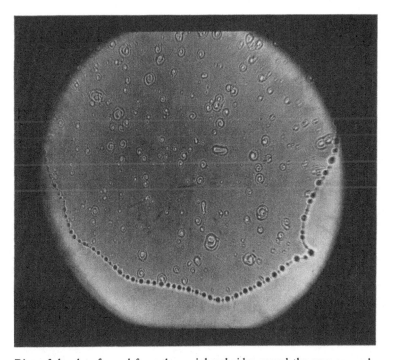

Ring of droplets formed from the peripheral ridge round the area covered
when a drop of aqueous solution has spread on mercury. Many breaks are
developing in the film, giving ridges that subsequently form rings of smaller
droplets.

The upper curve is for the spreading of a drop having a very minute trace of salt present and shows a more rapid stage at first, presumably until the concentration of ions falls very nearly to that in pure water.

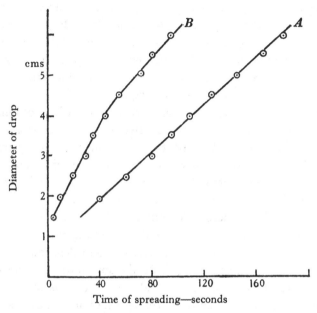

Fig. 12. Rate of spreading over mercury. *A*, distilled water; *B*, tap-water

Fig. 11 *C* shows the start of spreading of distilled water and indicates the relative rates for distilled water and dilute acid. The initial contraction is a recovery from the flattening due to falling on the mercury. In the cases *A* and *B* there is no such contraction shown, the ample supply of ions available enabling the drop to carry on from the area to which it was spread by falling on to the mercury.

Drops more concentrated than *B* show a definite but slight acceleration in the rate of growth of the diameter as the drop spreads. The fact that a drop of aqueous solution spreads over mercury at a rate which is very nearly constant from start to finish suggests that processes occurring at the periphery rather than over the interface may be the deciding factor. The steady speed is typical of motions due to a steady force and opposed by a viscous resistance, as when a body falls through a viscous medium.

3. Spreading controlled by Electric Current across the Interface

The well-known effect of polarization in changing the interfacial tension between mercury and solutions, as shown in the capillary electrometer, suggests that spreading might be controlled by an electromotive force applied across the interface. Experiment confirms this suggestion and also raises the question of the significance of measurements of the interfacial tension of water and mercury.

If one terminal (e.g. a platinum wire) be dipped into the mercury and another into a drop of water or aqueous solution placed on the surface of the mercury, the application of a small electromotive force between the terminals will accelerate the spreading when the mercury is positive and retard it when the mercury is negative.

(1) *Distilled water.* If the rate of growth in area of the spreading drop be measured both with the circuit open and with it closed, we find that the passage of 16 microcoulombs (measured by an ammeter in the circuit) across the interface will cause an extra square centimetre of surface to be covered. Sixteen microcoulombs are very nearly equivalent to 10^{14} electrons or univalent ions, a figure already obtained in connection with the spreading of dilute solutions of acid. If the current is reversed, the spreading is also reversed and the water is collected to a drop again, though in this case the connection between change of area and charge is not so definite.

If the potential applied in the reverse direction (i.e. mercury negative) is further increased beyond that needed to collect the water into a drop round the terminal, then the drop begins to spread again. If the circuit is opened after the drop has been spread to a disc in this manner, the water immediately retreats into a drop again, after which it will begin to spread in the ordinary way.

(2) *Very dilute solutions of acids.* The natural spreading is far too rapid for any acceleration due to electric current to be detected. With the polarity reversed a drop may be held from spreading or even forced to spread to any desired size by forcing across the interface sufficient current to keep the ions there against their tendency to diffuse away. On opening the circuit the acid flashes back to a drop and then spreads again in the ordinary way for acids.

Thus the driving of negative ions from solution to mercury causes stable spreading and their removal reverses this. The driving of positive ions on to the mercury produces a forced unstable spreading which is reversed immediately the forcing is removed.

(3) *Alkaline solutions* (hydroxides of sodium, ammonium, etc.). Drops of these non-spreading solutions may be forced to spread stably with the mercury positive. This spreading is usually accompanied by visible deposition, probably of a hydroxide of mercury, at the interface.

With the mercury negative a forced unstable spreading may be produced, the solution collecting again to a non-spreading drop as soon as the circuit is open.

(4) *Comparison with capillary-electric phenomena.* The effects described above could have been predicted from experiments with the capillary electrometer, though such experiments have usually been carried out with much more concentrated solutions. Many phenomena indicate that when two substances are put into contact a transference of electricity occurs across the common boundary. This results in the building up, at or near the interface, of a double layer of electric charges which opposes a further transfer of ions or electrons. Thus, for instance, the interface between a solution and mercury may be looked on as having a layer of positive charges on one side of the interface and of negative on the other. In whichever sense this quasi-condenser is charged, its effect is to reduce the free energy of the interface, or lower the interfacial tension. We may regard this reduction as being brought about either because the attraction between the positive and negative sheets of charge increases the affinities of the liquids at the interface, or because the repulsion between the like charges in either sheet gives forces in opposition to those producing the phenomenon of surface tension (as an electric charge will cause a soap bubble to expand). From either point of view we should expect the reduction from the maximum value to be proportional to the square of the charge density, as is shown by the parabolic form of the ideal electro-capillary curve. (Freundlich, *Colloid and Capillary Chemistry*, ch. v.)

When the electric current aids spreading, e.g. in the case of water, it speeds up the sorting out of ions to provide the necessary double layer at the interface. On reversing the current the double layer is discharged, the complete discharge (corresponding to the maximum of the electro-capillary curve) is attained when the spreading is reversed and the solution collected again to a drop. The condition in which there is no charge on the double layer corresponds to the greatest value possible for the interfacial tension.

The subsequent forced spreading corresponds to a forced charging of the double layer in an opposite sense to that which naturally occurs when a spreading solution is in contact with mercury.

4. The Interfacial Tension of Mercury and Pure Water

Values given in the tables for this quantity may range from 370 to 430 dynes/cm., giving of course a wide range of possible values for the spreading coefficient according to the values adopted, but a consideration of the phenomena of spreading indicates, first, that the spreading coefficient is probably very small, and, secondly, that different values for the interfacial tension are to be expected from different methods of measurement.

(1) *Antonow's rule.* Experiments described earlier show that the spreading of water on mercury may be completely stopped by traces of alkali in the water, or accelerated a hundred-fold by traces of acid. In neither case is the addition sufficient to affect the surface tension of the water to a measurable degree. The acid produces a slight lowering of the interfacial tension, thus making the spreading coefficient positive, while the alkali produces a rise in the interfacial tension and makes the spreading coefficient negative. That is, a very small change in γ_{AB} reverses the sign of the quantity

$$\gamma_B - \gamma_A - \gamma_{AB},$$

and so, at least very nearly,

$$\gamma_B - \gamma_A = \gamma_{AB}.$$

This is of course in agreement with Antonow's rule that for two mutually saturated liquids the interfacial tension is equal to the difference of their surface tensions. The generality of this rule is open to question [Ch. VII, 3, (5)], but it may be accepted as holding for mercury and water since they are not mutually soluble. The spreading of pure water over mercury is stopped if the air is saturated with water vapour. Whatever value is accepted for the surface tension of mercury, the interfacial tension against water must be about 72 dynes/cm. lower.

[It is sometimes stated that the addition of any substance to water lowers its interfacial tension against mercury, but as a matter of fact alkalis raise it, and for a concentrated solution of KOH the value of the interfacial tension against mercury may approach that of the surface tension of pure mercury.]

(2) *Pure water as a solution of ions.* Harkins obtained the value 375 dynes/cm. for the interfacial tension of water against mercury using the drop-weight method, the drops being formed slowly.

Gouy's value (*Ann. de Physique*, 6, 5, 1916) was 427 dynes/cm., obtained from measurements on a surface polarized so that the tension was a maximum. It was argued that this condition, which corresponds to the complete discharge of the double layer, should yield the true value for pure water against mercury. However, as we have just seen, the observations on electrically controlled spreading show that this maximum value of the interfacial tension occurs when water will not spread over and wet mercury, i.e. when the interfacial tension is greater than for the ordinary case of water in contact with mercury.

The method using polarization might be regarded as giving the interfacial tension for completely undissociated water and mercury; other methods should give a lower value because the mere wetting of the mercury by the water means an adsorption of 10^{14} ions/sq.cm. and a corresponding reduction in interfacial tension.

If the difference (52 ergs/cm.^{2}) between Harkins's and Gouy's values for the free energy of the water-mercury interface is regarded as the energy of the charged double layer, then the capacity of this layer is approximately 25 microfarads/cm.^{2}, and the potential difference between the layers is two-thirds of a volt.

The spreading of water on mercury, then, is probably not to be regarded as occurring between two pure liquids as is the case when benzene spreads on water. Rather we must consider that even conductivity water is a solution of ions, its rate of spreading depending on the number of these and their ability to reach the interface.

Ions are the 'active' constituents that control the spreading of water on mercury, just as (Ch. VI) a non-spreading paraffin may be caused to spread to a definite area and at varying rates over water by dissolving in the paraffin traces of 'active' materials which are adsorbed at the interface and lower its free energy.

CHAPTER VI

SPREADING ON WATER

1. Introduction

Among the earliest evidence of a limiting thickness for a continuous layer of material were the experiments of Miss Pockels and particularly the discovery of Lord Rayleigh that a drop of olive oil sufficient to give a layer 10^{-7} cm. ($1\mu\mu$) in thickness had practically no effect on the surface tension of water, while twice this amount would reduce the surface tension by 25 % and prevent the gyrations of particles of camphor on the surface. Rayleigh rightly judged that the sudden fall in surface tension corresponded to the completion of a single layer of molecules over the surface, and his work introduced by far the simplest means of determining the actual dimensions of certain types of molecule.

The development of methods of handling films on the surface of water, and particularly the technique of Langmuir and Adam, has brought the determination of molecular dimensions by measuring films spread on water to a high state of precision and has given information as to the properties of different parts of organic molecules and the forces between them. Later we shall give a brief summary of some of this work, an up-to-date account of which, with references to original sources, is to be found in the 1941 edition of Adam's *Physics and Chemistry of Surfaces*.

The general conditions for the spreading of one pure liquid on another have been stated (Ch. IV) and may be summed up in the statement that the adhesion between upper and lower liquid must exceed the cohesion of the upper one, if spreading is to occur.

Anything that lowers the interfacial tension between upper liquid and water increases the tendency for spreading to occur, while contamination of the water surface reduces this tendency.

2. Non-spreading Liquids

When experimental evidence disproved the belief that one liquid would always spread over another, the idea next put forward was that saturated compounds would not spread. Spreading and indeed all surface phenomena were to be explained in terms of 'chemical' rather than 'physical' forces, and only in unsaturated compounds could the molecules exert the forces necessary to produce spreading. It is now known that such hard and fast rules cannot be laid down. Nevertheless, the higher members of the series of saturated hydrocarbons (C_nH_{2n+2}) do form the most familiar group of nonspreading liquids. The new idea served to stress the fact that the range of the molecular forces involved in surface phenomena is very short. The forces fall off with distance much more rapidly, for instance, than gravitational forces. Though the force between two point charges of electricity obeys an inverse square law, that between two doublets falls off with the fourth power of the distance. In surface phenomena, however, we are usually concerned with the properties of, and forces between, *parts* of the different molecules. A simple law of force is not likely to hold, and the actual force between two contiguous molecules will generally depend upon their orientation with respect to each other.

3. Fatty Acids

(1) *Behaviour on the surface of water.* After Hardy's announcement that some liquids did not spread on water, the necessity for experimenting with pure substances rather than olive oil or mineral oils was recognized. The amount of oil needed for a monomolecular layer over a small area is so minute that it is usual to dissolve the oil in a volatile spreading agent such as benzene and administer the solution by a calibrated pipette. The benzene evaporates, leaving the molecules of oil floating on the water.

With a solution about o·oo1 N of a fatty acid such as stearic, $CH_3(CH_2)_{16}COOH$, or palmitic, $CH_3(CH_2)_{14}COOH$, and a clean water surface of 6 or 8 in. diameter, the first drop of solution spreads rapidly, showing film colours, and then evaporates. After a few drops have done this, the benzene spreads only to a circle smaller than the water surface. Each successive drop spreads to a smaller circle, and eventually a drop remains as a small lens as in the case of a drop of non-spreading paraffin on clean water. This occurs when the water is covered with a close-packed layer of molecules of acid. Simple calculation from the concentration and the size and number of drops used will show that approximately

$4·5 \times 10^{14}$ molecules of acid per sq.cm. have been applied to the surface of the water.

[The Ångström unit, 10^{-8} cm., is conveniently used in expressing the size of molecules, the unit of area being 1 sq. A. or 10^{-16} sq.cm.]

If different acids from the series are used, it is found that the spreading of benzene is stopped when the same number of molecules per sq.cm. are present on the surface of the water, irrespective of differences in size of the molecules of the various acids.

In acids of the series $CH_3(CH_2)_n COOH$ the groups COOH tend to make the molecule soluble in water and the early members of the series (formic, acetic, etc.) are in fact readily soluble. With increasing numbers of CH_2 groups in the molecule, however, the 'hydrophilic' head (COOH) of the molecule is no longer able to pull the hydrocarbon chain into the water, and in the close-packed layer of stearic acid the COOH groups are attached to the water and the hydrocarbon chains project from the surface, steeply, but not necessarily vertically, rather like the pile of a carpet. Higher numbers of the series simply have longer chains but do not occupy any greater area on the surface of the water.

This monolayer spread on the water presents an upper surface consisting only of the saturated CH_3 groups, and over this saturated layer the benzene is unable to spread. This is somewhat remarkable, since benzene can spread over and dissolve both stearic acid and paraffin. Evidently the COOH group is more strongly attracted to the water than is any part of the molecule of acid attracted by the benzene.

(2) *Molecular areas from spreading.* By using a dish, pipette and solution such that about 20 drops are required to prevent spreading, it is possible to obtain values for the number of molecules per sq.cm. which are consistent to within 1 or 2%. The author obtained 22·5 sq. A./molecule for stearic acid when benzene was prevented from spreading. Measurements on a similar film with a somewhat crude form of Langmuir balance (§4) suggested that the film contained some compressible impurity and showed that the benzene was prevented from spreading when the surface film was under a surface pressure of 10 dynes/cm. Tests with other substances indicated that in general benzene was unable to spread

against this compression. This is in fairly close agreement with the numerical value of the spreading coefficient of benzene on water, viz.

$$\gamma_A - \gamma_B - \gamma_{AB} = 73 - 29 - 35 = 9 \text{ ergs/cm.}^2 \text{ approximately.}$$

Tests were made on other monolayers (chiefly of malonates kindly supplied by Professor A. K. Macbeth) and it was found that when pure benzene would no longer spread over the water it often happened that a drop of 0·001 N solution of stearic acid in benzene would do so. At times sufficient stearic acid to form a complete monolayer of stearic acid could be added, in which case it was found that the original film would collapse completely under a compression of 18 dynes/cm. or less. With other films the stearic acid spread only to compress the monolayer to a definite smaller area, i.e. that occupied by the film at about 18 dynes/cm. compression. Thus the solution of stearic acid in benzene is able to spread over water against a retarding force of up to 18 dynes/cm. The simplicity and speed of this method of observation makes it of some use for measuring molecular size where the film will only persist for a minute or two owing to the solubility of its molecules in water.

Transue, Washburn and Kahler (*Jour. Am. Chem. Soc.* **64**, 274, 1942) have developed an essentially similar method for measuring the spreading pressure of volatile liquids. A drop of the liquid is placed on a water surface covered by a stable monolayer and the pressure exerted by the liquid is transmitted to the Langmuir barrier by the monolayer.

The term 'piston oil' is used for an oil having a definite spreading pressure which is used in order to maintain a constant pressure against a film on water. It functions as a flexible barrier, expanding or contracting as required.

4. Measurements on Monomolecular Films on Water

(1) *The Langmuir-Adam balance.* The technique developed by Pockels, Devaux, Labrouste, and particularly Langmuir and N. K. Adam, has enabled molecular films on water to be studied in great detail. No more than a brief summary of this most important work can be given here.

Experiments are made using a rectangular dish having a wide, truly flat edge and coated with paraffin wax (Fig. 13). As water does not wet paraffin, the dish may be filled so that the surface of the water is above the level of the edge of the dish. A strip of glass, longer than the width of the dish, and coated with paraffin, can be caused to slide along, sweeping ahead of it any insoluble molecules floating on the surface, thus confining any contamination to the area *A* and leaving the rest of the surface clean.

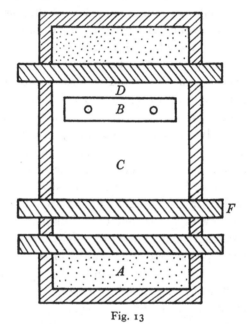

Fig. 13

A light floating barrier *B*, of paraffined paper or metal, reaches nearly to the edges of the dish and is prevented from moving along by the arms of a torsion balance mounted above the dish. These arms may be pointers which dip through the holes in *B*, and the total force tending to push *B* along the surface can be measured by the balance.

Naturally molecules floating on the surface *C* will produce a pressure tending to make some of them escape past the ends of the barrier *B* to the clear area *D*. This may be detected by the streaming through the gaps of talc dusted on to the surface. Langmuir prevented this leak by means of two carefully adjusted jets of air which played on the water surface in the gaps at the end of *B*. Adam joins *B* to the edges of the dish by thin ribbons of metallic leaf coated with paraffin, while others have used oiled threads to prevent leakage through the gaps. A known amount of the substance to be tested is placed on C, and by moving F, the 'surface pressure' exerted by the film when compressed to different areas can be measured, an experiment in two dimensions strictly analogous to finding the pv isothermal for a gas or other substance in three dimensions.

In expert hands 'pressures' as low as 0·01 dyne/cm. have been measured.

(2) *Types of film.* The pioneer work of Langmuir established the existence of films consisting of single layers of orientated molecules. More recent work has enabled different types of film to be recognized, these types corresponding in a general way to the various states of a substance in bulk.

(*a*) *Gaseous films.* Where the number of molecules is too small to form a continuous layer over the surface of the water we might expect one of two possibilities.

(i) The molecules distribute themselves uniformly over the surface like a two-dimensioned gas possessing the same average energy per degree of freedom per molecule as the water on which they float and exerting in consequence a pressure against the confining barriers.

[Just as n molecules of gas enclosed in a cube of 1 centimetre edge exert by their impacts a force $n.k.T$ on each face of the cube in virtue of their average energy $\frac{3}{2}kT$, so n molecules floating on 1 sq.cm. of the surface of water exert a 'surface pressure' on each boundary given by

$$F = n.k.T$$

in virtue of their average energy of kT, i.e. of $\frac{1}{2}kT$ for each degree of freedom.

If we write a for $1/n$, the area per molecule, the equation becomes

$$F.a = k.T. \tag{1}$$

$k = 1·37 \times 10^{-16}$ ergs/cm., so that expressing a in sq. A. (or 10^{-16} sq.cm.)

$$F.a = 1·37T.]$$

For ordinary room temperatures (*c.* 290° Absolute) the product $1·37T$ is very nearly 400 and films have been measured in which the product $F . a$ is very near this value, indicating a perfect gas in two dimensions. This will occur only for large values of a, where the actual size of the molecule is negligible compared with the space it occupies on the surface (cf. gases in three dimensions). More usually these two-dimensional gases are imperfect [4 (2) c].

In gaseous films the molecules almost certainly do not float with their heads in the water and the hydrocarbon chains vertical like masts. They must take up positions of minimum energy and, owing to adhesion between hydrocarbon chain and water, this means that the molecules will lie flat on the water.

(ii) The second possibility is that some of the molecules may be collected into patches (liquid or solid) and the remainder spread uniformly over the surface as a two-dimensional vapour, exerting a definite 'vapour pressure' on the barrier. Reducing the available area does not increase the pressure but causes more of the vapour to condense, just as with a saturated vapour in three dimensions.

(b) *Close-packed or condensed films.* In these the molecules are in close contact, with their hydrocarbon chains either vertical or steeply orientated, and the film may be either liquid or solid. If talc is dusted on the surface and gently blown, it can be seen whether the surface can flow as a fluid or whether the film is solid, moving in rigid patches or floes when the force is great enough.

In members of a given series, those molecules having the longer chains tend to give solid films owing to cohesion between the hydrocarbon chains. Such solid monolayers may withstand a compression of 40 dynes/cm. or more without collapsing, and they exhibit a compressibility of the same order as that of the substance in bulk.

(c) *Expanded films.* Labrouste (*Ann. de Physique*, **14**, 164, 1920) recorded a change in the area of monomolecular films, occurring at a fairly definite temperature for a given compound and apparently corresponding to a change in state of the film.

More accurate measurements by Adam show that if the film is maintained under a constant small pressure the change sets in at a definite temperature which depends on the pressure. The area-temperature curves at any pressure have the same general characteristics as volume-temperature curves for a change of state, e.g. of water into steam, though the change is not quite isothermal, there being a range of several degrees between the beginning and end of the transition from condensed to expanded films. Similarly the isothermal showing pressure against area per molecule bears a general resemblance to the isothermal pv curve for the conversion of water to steam (Fig. 14).

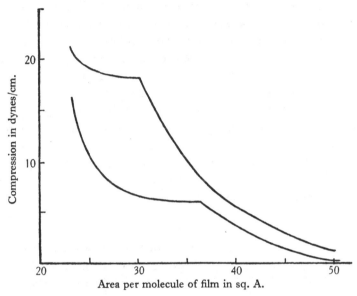

Fig. 14. Shapes of the $F.a$ curve (isothermal) at different temperatures for the change from gaseous films at large areas, to condensed films at high compressions.

The area per molecule in these expanded films does not correspond to any definite orientation. They are of interest here because Langmuir worked out the theory of the most common type of these films largely from experiments carried out by Dr Katharine Blodgett on the spreading of oils on water (*Jour. Chem. Phys.* **1**, 756, 1933). Langmuir has called these monolayers* 'duplex' films and regards them as having an upper and lower surface which may be treated as two independent surfaces. The upper surface, composed of the CH_3 groups, behaves as a free liquid surface, having, for example, the same surface tension as a hydrocarbon liquid in

---

* Harkins, however, applies the term 'duplex' to a film, thicker than a monolayer, but thin enough for gravitational forces to be ignored.

bulk. The lower surface of the film is regarded as an interface in which the water-soluble ends move about like the molecules of a gaseous film but are so close together that their $F.a$ equation becomes

$$F(a - a_0) = k.T, \qquad (2)$$

where a_0 represents the diminution in free area due to the molecules themselves. [The a_0 corresponds to van der Waals's b but no correction for force between the molecules corresponding to the a/v^2 of van der Waals is made in the two-dimensional case.]

In such duplex films the increased thermal agitation (which has brought about the transition from a condensed film) prevents the chains from being closely packed and they are in random movement like the molecules of a liquid. Nevertheless, there is sufficient coherence between the chains to prevent them separating like the molecules of a gas. Such a view of course entails the assumption of a high degree of flexibility in the hydrocarbon chains.

According to Langmuir's theory such a duplex film will have a spreading force F given by the sum

$$F_0 + F_{AB},$$

where F_0 is the spreading coefficient of the hydrocarbon on water (usually negative) and $F_{AB} = kT/(a - a_0)$, the gaseous pressure due to the heads of the molecules at the interface (as in equation (2)).

Whence $\qquad\qquad F = F_0 + F_{AB}$

and $\qquad\qquad F - F_0 = kT/(a - a_0).$

This equation fits the $F.a$ values for expanded films until the point is reached where some of the molecules become close packed in groups, somewhat analogous to the condensation to a mist in the case of vapours near the critical point.

5. Test of Homogeneity of Films

The assumption that a film, whether monomolecular or thicker, is uniform over its area may be tested optically, or by measurement of surface-potential.

(1) *Optical methods.* Unspread particles may be seen by microscopic examination, particularly with the use of dark ground illumination as in the ultra-microscope.

By the application of Drude's theory of the reflection of polarized light from thin films attempts have been made to measure the thickness of films (Tronstad and Feachem, *Proc. Roy. Soc.* A, **145**, 115, 1934, etc.). While the value of these methods for determining the thickness of adsorbed layers is perhaps doubtful, the method may be employed to detect variations from point to point in the structure of the film.

(2) *Method of surface-potential measurement.* In the method described by Zisman (*Rev. Sci. Inst.* **3**, 367, 1932 and *Physics*, Jan. 1933) a vibrating plate, parallel and close to the surface, forms one plate of a condenser and the surface of the liquid forms the other. A potential difference between the plates gives rise to an alternating current owing to the capacity variations, and this current is amplified and gives a note in the earphones. The difference in potential between plate surfaces is found from the potential that must be applied to the vibrating plate in order to silence the note. This enables the uniform surface potential over a small area to be determined with amazing precision. With a non-homogeneous surface, however, the note cannot be silenced. The vibrating plate, having a carefully aged surface, is maintained in vibration by mechanical means, at a frequency in the audible range.

The method of the probe or air-electrode enables areas to be examined from point to point (Schulman and Rideal, *Proc. Roy. Soc.* A, **130**, 259, 1931). A pointer coated with polonium is placed very near the surface and acquires the potential of the surface by reason of the ionization of the air. The potential is then measured by an electrometer. By traversing the film with the electrode a change in nature of the surface at any point can be detected. Measurements of surface potential together with those of pressure and area per molecule are now used in experiments on the orientation and electrical moments of molecules in films.

6. Solutions of Active Substances in Non-spreading Paraffins

In Ch. V we saw that the addition of an acid to pure water caused a drop to spread rapidly on the surface of mercury to cover a definite area, proportional to the number of molecules of acid in the drop but greater than that required for a simple monolayer. Similarly the addition of a minute quantity of a fatty acid to a paraffin will cause the paraffin to spread to a definite area on the surface of water. The active molecules diffuse to the interface, so reducing the interfacial tension γ_{AB} and causing a positive spreading co-efficient. In the early stages of the spreading the relation $dA/dt \propto A$ is also found to hold.

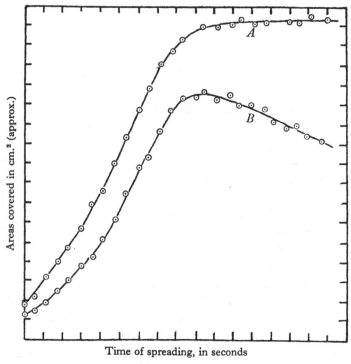

Fig. 15. Spreading of paraffin containing 0·02 % of stearic acid on water containing both sodium and calcium ions. *A*, area covered by film; *B*, area of visible drop of paraffin (see Plate II)

Langmuir (*Jour. Franklin Inst.* **218**, 143, 1934) found that on alkaline water containing only sodium ions a drop of a dilute solution of stearic acid in petrolatum spread to a final area of 88 sq. A. per molecule of stearic acid. Traces of calcium present in the water caused the area per molecule to fall to 22 sq. A. (approximately that for a close-packed film) and the film at the interface became a two-dimensional solid. On 0·01 N solution of HCl a given solution of stearic acid spreads to a much smaller area,

PLATE II

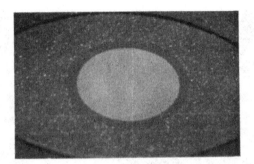

Paraffin with 0·02 % stearic acid spreading over water which contains calcium ions. A layer of stearic acid or stearate shed from the paraffin spreads ahead of it, clearing an annular space of the aluminium powder with which the surface of the water was dusted.

PLATE III *facing page* 63

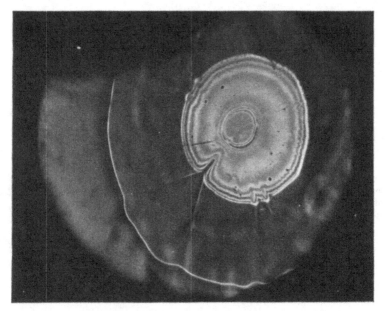

A drop of benzene spreading over a slightly contaminated surface of water. Newton's rings show the limit of the visible film. The sharp white line further out apparently marks the limit of an invisible film.

probably due to solution of the adsorbed molecules in the petrolatum.

Mercer (*Proc. Phys. Soc.* 51, 561, 1939) observed the spreading of a solution of stearic acid in paraffin oil on water which was made alkaline with NaOH and which also contained calcium ions. If aluminium powder is dusted over the surface of the water on which the drop is spreading, an annular clear area presently appears round the spreading oil. This is due to the aluminium being pushed aside by a (probably monomolecular) layer of stearic acid (or calcium stearate) which is being shed on to the water surface by the spreading drop (Plate II).

Apparently the active molecules are carried out by the water in the usual method of spreading, but the adhesion between the hydrocarbon chains and paraffin is so slight that the drop of paraffin is continually losing its hold on the spreading monolayer and contracting (Fig. 15). Eventually the paraffin retreats to one or more lenses on the stable monolayer.

Mercer finds this shedding of stearic acid from the drop is very slight if only sodium or potassium ions are present in the water. The spreading of an invisible film ahead of the visible spreading drop occurs even in cases of rapid spreading (cf. Plate III). McBain has observed the effect in the spreading of pure oleic acid on water.

Zisman (*Jour. Chem. Phys.* 9, pp. 534, 729 and 789, 1941) has studied the spreading of paraffin solutions of both ionized and unionized spreaders and finds that the effects observed by Mercer are of general occurrence. The rate of spreading depends on the nature and concentration of the active substance and the viscosity of the paraffin, being faster for shorter carbon chains and for less viscous paraffins. The shedding, or two-dimensional evaporation, of the spreader from the paraffin is also faster the shorter the chain and has practically ceased for chains containing 20 carbon atoms. Molecular agitation is not able to free the long carbon chains from the paraffin and no monolayer is spread from the disc. Zisman finds that when the spreader is not ionized no spreading occurs until a definite minimum concentration is reached. The life of unionized molecules in the interface is short and spreading occurs only while the rate of adsorption of molecules at the interface exceeds their rate of release.

7. Nature of Interfacial Films and Monolayers

(1) *Effect of metallic ions.* From the time of the earliest precise work on films observers noticed effects that they could not assign to any definite cause, such as changes in area per molecule with the time the water stood in the trough. Subsequently it was shown that an extremely small concentration of metallic ions, especially those of higher valency, produced fundamental changes in the nature of surface films and interfaces. Langmuir showed that a film of stearic acid, which was always liquid when on pure water made alkaline with KOH, was converted to a solid film by the presence of one part in 10^8 of calcium ions in the water (*Science*, **84**, 379, 1936). Others have shown the need for extreme care in avoiding metallic ions, particularly copper, and the practice now is to use troughs of hard glass or quartz. Zisman (*loc. cit.*) studied the interfacial film between water and a paraffin solution of fatty acids. For chains of 16 or more carbon atoms the interfacial film was made rigid by a concentration of 10^{-4}N of salts of Ca, Pb, Cu, Fe. For chains of 14 or 15 carbon atoms the films were made rigid by ions of higher valency while still shorter chains did not give rigid films under the action of metallic ions.

(2) *Complexes in monolayers.* The greatest area to which the paraffin disc spreads may be considerably bigger than that required if all the active molecules formed a monolayer on the surface. The interface apparently is a mixture of active molecules with those of the hydrocarbon. When the interface is caused to contract under increasing applied pressure, evidence is obtained that the mixed film is not a simple mixture of polar and non-polar molecules, but consists of 2-dimensional associations or 'complexes' in which active and inactive molecules are present in definite ratios 1 : 1, or 1 : 2, etc. Joly considers that the existence of such complexes in monolayers is established if, during the collapse of the film under pressure, there are definite ranges of pressure under which the film is stable. He found that with sodium cetyl alcohol either a 1 : 1 or a 1 : 3 complex was formed according to the area of surface available (*Nature*, **158**, 26, 1946).

In the Liversidge lecture for 1945 (*Jour. Chem. Soc.* p. 423) Rideal points out that the chemistry of reactions in monolayers is probably of great biological importance. It has already been established that fatty acids, chloresterol, etc. form complexes with monolayers of proteins and much of the work on interfacial films now in progress is directed towards possible biological applications.

8. Multilayer Films on Solids

Dr Blodgett (*Jour. Am. Chem. Soc.* **57**, 1007, 1935) has developed a method by which successive monolayers can be deposited on a glass slide. Stearic acid is spread on water containing calcium, the film being kept under constant compression by a drop of a spreading oil placed on the water surface. A glass slide raised through the surface acquires a coating of orientated molecules of calcium stearate. By alternately raising and lowering the slide successive folds of the film may be deposited on the glass, the heads being alternately inward and outward in successive layers. A variation of technique enables films all having the same orientation to be deposited. By the reflection of monochromatic light Dr Blodgett (*Phys. Rev.* **51**, 964, 1937) has been able to measure the thickness of these multilayers to within a single layer and also to detect variations in their composition, as e.g. when a stearate film is treated with benzene to remove the stearic acid molecules and leave a *skeleton* film of neutral stearate. Protein layers have been deposited over skeleton films of barium stearate in this way and their permeability to vapour studied by noting the time for the colour to change, indicating that the vapour had penetrated the protein and filled the holes in the skeleton film of stearate.

Langmuir (*Science*, **87**, 493, 1938) has put forward the view that molecules in these monolayers, whether of acids or proteins, can suddenly turn end for end in the film, the hydrophilic and hydrophobic ends changing places without the molecules leaving the film or the film as a whole peeling off. The evidence is based on the changes in behaviour towards water (e.g. angle of contact) which occur in these deposited films. Langmuir suggests that such overturning of surface layers may have a bearing on the propagation of nerve impulses.

The most recent application of this process of applying a succession of monomolecular layers to glass is in the production of a nonreflecting or invisible surface on glass (*Phys. Rev.* 15 Feb. 1939).

Dr Blodgett applies to the surface of the glass sufficient layers of cadmium arachidate (about 45 monolayers) to give a film having an optical thickness of a quarter wave-length for the middle of the visible spectrum. Thus light reflected from the first surface of the film is out of phase with that reflected from the interface between film and glass. The result is an almost complete interference between these beams, thus eliminating reflection from the glass. An extension of this process may provide the desired 'invisible glass' for shop windows, and it obviously has many applications, such as eliminating glare from the glass covers of instruments, from spectacle lenses, etc. The absence of a reflected beam from the surface also means a corresponding increase in the transmission of light through the glass, thus improving the performance of lenses, etc. The reduction of reflection in certain types of optical instruments by 'blooming' with a solid layer having a refractive index midway between that of air and glass, and an optical thickness of one-quarter wave-length, is now almost standard practice.

9. Paraffin caused to Spread by Irradiation

If a drop of non-spreading paraffin is placed on a dish of water in sunlight, a certain amount of spreading is caused. When the oil is irradiated with γ-rays, X-rays, or ultra-violet light, the result is a spreading similar to that produced by traces of fatty acid.

Observations on this activation are recorded by Stenstrom and Vigness (*Jour. Chem. Phys.* **5**, 298, 1937) and Allen, Grant and Burdon (*Trans. Far. Soc.* **33**, 1531, 1937). The nature of the activation was not discovered. Every effort was made to secure pure paraffins in order to be certain that the action of the radiation was not simply to produce a change in an impurity in the paraffin. The irradiation with ultra-violet light was carried out with oxygen excluded, so that the effect does not appear to be due to a formation of fatty acids from the paraffin. Paraffin will spread even when kept for months after it has been irradiated, so that the effect appears to be permanent.

Fig. 16 was obtained by Allen by plotting the diameter of a spreading drop against the square of the time. A different plot would show that in the early stages (curved part of the graph) the relation $dA/dt \propto A$ holds fairly closely. The straight part, however, shows that over this range the edge of the spreading drop advances with uniform acceleration (cf. the expression for a falling body, $s = \frac{1}{2}gt^2$)—a most unexpected law to hold for a viscous film spreading over water. Though the curve could be repeated accurately, it is unlikely to afford a basis for any simple theory as to the nature of

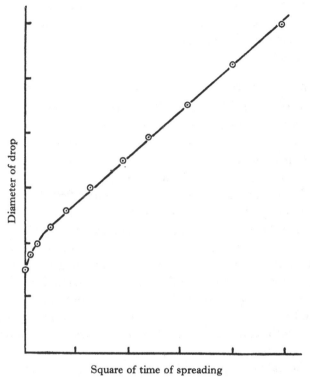

Square of time of spreading

Fig. 16. The rectilinear portion of the curve indicates spreading with constant acceleration

spreading and the forces involved. Observations on the speed of spreading often surprise one by the precision with which they may be repeated. The worker soon finds that a linear graph does not mean that a simple theory of the phenomena can be given. (Cf. Landt and Volmer, *Zeit. Phys. Chem.* **122**, 398, 1926.)

Stenstrom and Vigness have published an account of further work on the irradiation of heavy mineral oils (*Jour. Phys. Chem.* **43**, 531, 1939). They find that ultra-violet light of wave-length less than 2800 A. produces changes in heavy hydrocarbons. When irradiated in vacuum a small fraction of the molecules is rendered unsaturated and there is evolved an explosive gas, probably consisting of hydrogen and short-chain hydrocarbons. In contrast with the observations of Allen, they find that the irradiation in vacuum does not cause the paraffin to spread on water. When oxygen is present the irradiation results in oxidation with the formation of unsaturated acids, about one-tenth of the unsaturated molecules becoming oxidized. The spreading on water is due chiefly to the acids which are formed.

10. Spreading on a partially Contaminated Surface

Since the time of Osborne Reynolds observations have been recorded from time to time on phenomena which might be classed under this heading. Several writers have commented on the sharp line of division seen between parts of varying degrees of contamination on the surface of flowing water.

(1) *Visible limit of a monolayer.* Labrouste first demonstrated the possibility of seeing a monolayer (*Ann. de Physique*, 14, 164). Water in a shallow porcelain dish is illuminated at a glancing angle and on it is placed sufficient fatty acid to give a liquid monolayer over part of the surface. On blowing gently from one corner the monolayer is blown to one end of the dish and its limit is shown by a sharp line of light projected on the bottom of the dish. Adding more oil reduces the area which can be cleared of film by blowing over the surface, thus indicating that it is the limit of the film which is made visible.

(2) *Benzene on water.* A drop of benzene on a clean water surface apparently gives rise to a ridge which flashes across the surface. On a surface partially covered with a monolayer of olive oil the ridge or dividing line can easily be photographed. In Plate III the Newton's rings give a contour map of the visible benzene film, while some distance farther out is the sharp line presumably marking the limit to which an invisible film of benzene has spread.

(3) *Behaviour of aniline on water.* The late Professor Edser recommended to the writer the observation of aniline on water. A drop of aniline placed on the surface of water may either spread rapidly with streaming movements and disappear in a few seconds, leaving practically no contamination, or it may break through the

surface and sink to the bottom of the dish, where it remains as a spherical drop with apparently no reaction with the water.

If drops of aniline are placed on the surface of water which is progressively contaminated, some remarkable effects can be seen. The drops of aniline move about and are often distorted by rapid lifelike movements, dividing into smaller drops that become circular if quiescent for a moment, then again taking up rapid irregular contortions. Two or three drops placed on water in a small transparent dish enable the motion to be projected on a screen and give a fascinating display, suggestive of the part that surface variations can play in the movements of forms of life in a fluid medium. When there is sufficient contamination on the surface to form nearly a complete monolayer, the drops of aniline placed on the surface spread to discs of about 1 cm. diameter and after remaining quiescent for a few seconds each drop gives a violent twitch. The drops suddenly expand, then contract, and then settle down again for a second or two. This form of 'Kipp-oscillation' persists until the drops finally disappear by solution. Three or four drops on a small area generally keep time with their twitching, possibly because the disturbance from one drop sets off the others even if they are not quite ready to expand. This simultaneity is itself remarkable enough, since generally if a number of lenses of oil are endeavouring to expand against a film of contamination any expansion of one drop will transmit a pressure making the other drops contract.

When the drops are more widely separated on a large expanse of water, the individual drops appear to oscillate independently of each other.

If one starts with a clean surface of water, changes in the behaviour of drops of aniline will occur as the water becomes more nearly saturated with aniline, the stages being much like those described. On saturated water the aniline remains at rest.

Somewhat similar periodic variations in the surface of mercury drops in contact with potassium bichromate solutions have long been known (e.g. the so-called 'mercury heart').

11. Action of Oil on Surface Waves on Water

No chapter on oil films would be complete without reference to the effect of oil in calming the sea. Though a good deal of experimental work relating to this subject has been done over a long period, there is after all a wide gap between conditions holding in the experimental trough and those obtaining on an actual occasion when oil is used to permit the launching of a boat, and it is not surprising that differences of opinion still exist. The latter conditions are certainly not conducive to scientific observation. Directly contradictory statements can be found on the question of whether the oil reduces the amplitude of large waves, and also whether the wind has more or less purchase on a lubricated water surface than on a clean one (cf. Merigoux, *Compt. Rend.* **205**, 115, 1937 and Adam, *Physics and Chemistry of Surfaces*, p. 105, 1938). It seems improbable that any action by the oil could dissipate enough energy to make an appreciable difference to the amplitude of large waves in the time which they take to traverse a few chains of oil patch. The function of the oil is rather to reduce the immediate effect of the wind on the water at the spot. An active oil, giving a strongly adsorbed monolayer, is probably best, and the action is that of damping out ripples, or preventing their formation. In the absence of oil the wind is continually producing ripples which rapidly grow to a size where they can produce irregular breaking of the big waves. On the expanding and contracting surface of the water the monolayer would, by a kind of surface viscosity, dissipate the energy of these incipient ripples. The transmission of energy from the surface to lower layers is almost certainly an important factor, as it is in the rapid damping of waves by rain when the falling drops carry water from the surface to lower layers.

12. Influence of Surface Layers on Evaporation

Stable monolayers on the surface of water have no measurable effect on the total rate of loss of water by evaporation. This is to be expected since the measured rate of loss is never more than a small fraction of the theoretical rate at which molecules leave the surface, as calculated from the maximum vapour pressure for water.

A reduction of evaporation from water is important in some industries and during the war attention was devoted to the possibility of reducing evaporation from exposed water surfaces in arid regions.

Powell (*Trans. Far. Soc.* 39, 311, 1943) and Heymann and Yoffe (*Ibid.* 38, 408, and 39, 217) experimented with thick films of oil, the latter finding that a film of the order of 0·01 mm. thick caused a reduction of sixty per cent or more in the loss by evaporation from still water. Heymann has investigated the persistence of these films and found that when a substance such as polymerized oleic acid was used as a spreading agent the film of heavy oil was remarkably stable, evaporation rates through such a film being observed over a period of 18 months. The polymerized spreader gave a rigid interfacial film over which the hydrocarbon remained as a thick film, instead of contracting to lenses on a monolayer as is the case when simple fatty acids are the spreading agents.

CHAPTER VII

LIQUIDS ON THE SURFACE OF SOLIDS

1. General Remarks

As we have already seen (Ch. IV, 4), one result of the immobility of the molecules in the surface of the solid is that just those factors which promote spreading over the surface of a liquid may retard the process on the surface of a solid. Active molecules on the surface of water are spread by the diffusion of the water molecules. When placed on the surface of a solid the active molecules simply become anchored by attachment to the surface and so are incapable of further spreading.

Generally speaking, the surface of a solid will differ from point to point both in its structure and in the resultant fields of force and free energy. Even the most highly polished surface will not be comparable in uniformity of level and structure with the surface of a liquid.

Most solid surfaces which have been exposed to air for any appreciable time will be covered more or less completely with a film of contamination which will change the free energy of the surface and consequently alter its behaviour in regard to the spreading of liquids over it. The spontaneous spreading of a drop of liquid to form a circular patch on a solid is seldom observed, and even on carefully cleaned platinum in air, drops of water or of aqueous solutions do not exhibit phenomena similar to those observed on mercury.

Nevertheless, the behaviour of liquids with respect to solid surfaces is of the greatest technical importance. Owing to the difficulty in obtaining clean, easily renewable, solid surfaces it is only in the last few years that appreciable progress has been made in the study of liquids on the surface of solids.

In the industrial processes which involve the behaviour of liquids with respect to the surface of solids it is frequently a matter of indifference whether the liquid will spread spontaneously over the surface or not; the fundamental requirement in some cases (e.g. painting and spraying) is that the liquid, once spread, shall maintain a stable layer; in other cases (e.g. the flotation of minerals) it is fundamental that the solid surface shall *not* retain a stable layer of liquid over it, but that a bubble of air shall be able to make contact with the surface of the solid and displace the liquid from part of the surface (§4).

Active substances added to a liquid in order to secure the stability of a layer and prevent it retreating from the surface over which it has been spread are called wetting agents (§6).

Only in the case of diffusion through capillaries does the natural spreading of liquids over solids play much part, and even here mechanical agitation is used where possible in order to bring about contact of solid and liquid surfaces.

2. Lubrication

(1) *Complete lubrication: static friction absent.* The lubricated surfaces are kept apart by an appreciable layer of lubricant. If this is to be achieved, the lubricant must adhere to the moving surface and a sufficient layer must be dragged forward against the pressure to hold the bearing surfaces apart. The more viscous the lubricant the more it will be dragged forward by the layer adhering to the moving surface of the bearing. The faster the movement the less viscous need be the oil to maintain lubrication, since the higher velocity gradient across the layers of oil will produce a greater tangential drag to pull the oil between the surfaces.

Woog pointed out that in clock-oil an active constituent is necessary so that its molecules may adhere and form a barrier preventing the saturated components of the oil from spreading and disappearing from between the bearing surfaces. The superiority of animal oils over mineral oils depends largely on the presence of such active constituents which give adherence of lubricant to metal.

In the lubrication of a shaft the process by which the oil film is maintained is somewhat akin to the spreading of paraffin containing a fatty acid over the surface of water (Ch. VI, 6), the adhering film being carried forward by the moving surface of the shaft instead of by diffusion in the surface layer of the water. In some oils the adhesion to the surface of the bearings is increased by the addition of about 1 % of fatty acid, the molecules of which adhere at the interface and form a strongly held layer. This layer, carried forward by the motion of the shaft, drags a thicker film of lubricant with it. Experiment shows that if bearing surfaces are driven at such speed and under such pressure that lubrication breaks down when a pure mineral oil is used, then the addition of a trace of active material, e.g. stearic acid, to the oil will enable it to maintain its lubricating film between the surfaces. The addition of the fatty acid in effect permits us to use a less viscous oil than would otherwise be necessary under the conditions in question. Once lubrication is attained, greater viscosity simply means greater fluid friction.

Bowden, Leben and Tabor (*Trans. Far. Soc.* **35**, 900, 1939) report that by heating mineral oil, either when saturated with air or after ozone had passed through it, they had produced changes in the lubricating properties of the oil similar to those conferred by the addition of a fatty acid. The active compounds responsible for the change in lubrication are formed rapidly at higher temperatures (200° C. or over) and where the heat is applied to the oil on the bearing surfaces. Thus these changes will occur under working conditions such as exist in the cylinder of an engine. The improvement in properties of an oil under working conditions is only likely to be temporary, subsequent changes in the oil at high temperatures producing polymerization, gumming, etc.

In view of the observations on the effect of irradiation which were described in the previous chapter it would be of interest to test the effect of ultra-violet light on the lubricating properties of the paraffins.

(2) *Boundary lubrication: static friction present.* When the lubricant between the surfaces consists only of the films adhering to them the problem is called one of boundary lubrication. Static friction is present, and the surfaces are so close together that their nature as well as that of the lubricant plays a part in determining the magnitude of the friction. Hardy found that when surfaces were lubricated with only a monolayer of fatty acid the friction fell by a definite amount for each additional CH_2 group in the molecule of acid. This work on lubrication provided some of the earliest evidence for the existence of monolayers of orientated molecules. In such a (probably closely packed) film the active ends of the molecules are anchored to the solid and the longer the hydrocarbon

chain the farther the metal surfaces are separated, and the more friction is reduced.

These boundary layers are so firmly held that throughout a considerable range of temperature there is no variation in the friction (Hardy, *Phil. Trans.* A, **230**, 1, 1932, and Hardy and Nottage; *Proc. Roy. Soc.* A, **138**, 259, 1932, etc.). Where a temperature coefficient was found to exist in boundary lubrication it could be attributed to the presence of more than one molecular species in the lubricant, the variations occurring with the temperature being due to interchanges of the molecules adsorbed at the surface.

In boundary lubrication Hardy found that there was a latent period after the application of the lubricant before the friction reached a steady value, and he suggested that most of the wear on a bearing occurred in this latent period. The changes occurring during this latent period are possibly connected with the orientation of the molecules of the lubricant. The properties of these thin films of fluid between solid surfaces differ so much from those of the fluid in bulk that Hardy said they might be regarded as a fourth state of matter. It is uncertain to what extent the phenomena of boundary lubrication are due to single layers of orientated molecules adhering to the surfaces, but there is fairly general agreement that here we have the influence of the solid surface exerted beyond the first layer of molecules and that perhaps the orientation extends through a number of layers (cf. Hersey, *Theory of Lubrication*, 1936).

It is clear that the properties so far discussed, adherence and viscosity, are chiefly concerned with getting the film of lubricant between the surfaces. Its usefulness when there depends, in part at least, on another property which so far has not been sufficiently elucidated for a satisfactory definition to be given. This property is given the name 'oiliness' by many writers, and in technical circles it is called 'body'. Attempts to specify just in what the body, or oiliness, of an oil consists have not been very successful, but Hersey quotes Woog (*Compt. Rend.* **173**, 303, 1921) as classifying oiliness or body as a latent property of the lubricant brought into action by the proximity of the solid surface. Hersey (*loc. cit.*) says that oiliness refers to the frictional characteristics of the lubricant and that it is probably of greatest importance under conditions of high pressure and low speed such as obtain at the starting of railway carriages and heavy trucks.

(3) *Recent work.* This has been directed chiefly to the study of boundary layers. Dacus, Coleman and Roess (*Jour. App. Phys.* 15, 813, 1944) have investigated what they term the 'durability' of monolayers. A lubricant, such as barium stearate, is placed on a surface and rubbed down to a monolayer. The rim of a smooth wheel is caused to rub on this surface, exerting a constant pressure, and the durability is measured by the distance the rim moves before the layer breaks down and sticking or seizure occurs between the surfaces. Durability thus defined was found to be approximately proportional to the number of carbon atoms in the chain.

Bowden, Bastow and fellow-workers have studied the breakdown of boundary layers under various conditions. Metals were coated with layers of stearic acid and the friction measured at various temperatures. On noble metals (e.g. platinum) lubrication broke down at a temperature close to the melting point of the acid in bulk, whereas on base metals the lubrication persisted to a higher temperature owing to the formation of metallic stearates having higher melting points. Electron diffraction has been used to detect the disorientation of these boundary layers corresponding to the melting of the acid or soap and the findings from frictional measurements have generally been confirmed.

Use is now being made of radioactive isotopes in the study of boundary lubrication. A slider made of a metal containing a radioactive fraction is made to slide once over the second surface, either lubricated or bare. Photographic paper is then placed in contact with the second surface for some time. When subsequently developed, the relative amount of metal removed from the slider during its passage will be indicated by the relative development of parts of the exposed paper. Cf. Sackmann, Burwell and Irvine (*Jour. App. Phys.* 15, 459, 1944) and Gregory (C.S.I.R. (Australia) *Report No.* 54 *on Lubrication and Friction*, Jan. 1946), Spink and Gregory (*Nature*, 159, 403, 1947).

3. Angle of Contact

(1) *Its reality and importance.* The angle of contact between a liquid and solid is the only directly measurable quantity which

conveys any information about the free energy of the liquid-solid interface (Ch. IV, 4).

The non-uniformity of solid surfaces, and particularly the fact that the angle measured was frequently that between liquid and a film of contamination rather than that between liquid and the solid under observation, led to much uncertainty and inconsistency in early observations on contact angles. It is not many years since the opinion was expressed that contact angles, if indeed they were real, were at any rate too indefinite to be of value in discussing surface phenomena. Wark (*Principles of Flotation*, Aust. Inst. of Min. and Met. p. 115, 1938) says that measurements of contact angles made prior to 1931 are of historical interest only, since they were almost invariably made on contaminated surfaces. In actual flotation practice, however, it is the angle against a clean surface that matters, since the mineral is not exposed to contamination between the crushing and flotation processes (§4).

(2) *Hysteresis of the contact angle.* If a drop of water is placed on a plane surface of wax, or if a drop of mercury is placed on glass or steel and the surfaces are tilted till the drop moves, then the contact angle on the advancing edge is always greater than that on the receding edge (cf. the tail on a drop of contaminated mercury). Sulman (*Trans. Inst. Min. and Met.* 29, 44, 1919) called this effect the hysteresis of the contact angle. The advancing angle is that between the liquid and the unwetted surface of the solid, the receding angle that between the liquid and a surface that has been in contact with the liquid. These represent the limiting values which the contact angle can assume.

The experiments of Wark and his collaborators on specially clean and smooth mineral surfaces suggest that this hysteresis is a result of frictional resistance to movement of the liquid over the surface of the solid. If so, it is to be expected that a drop at rest should eventually settle down with the same angle of contact on all sides, since friction in a fluid cannot supply a static force (cf. Adam, *Physics and Chemistry of Surfaces*, p. 181).

The conditioning of the surface, particularly by a previous heating, will reduce the hysteresis in many cases, but in some instances it seems probable that the difference in contact angle persists. Hysteresis indicates that the passage of the drop has changed the nature of the surface over which it has passed, i.e. the *dry* surface of the solid from which the drop has receded differs from the surface that has not yet come into contact with the liquid. The contact of liquid and solid has increased the adhesion between them. In the relation for equilibrium,

$$\gamma_B = \gamma_{AB} + \gamma_A \cos \theta, \qquad [\text{Ch. iv, } 4 \, (1)];$$

if θ is decreased γ_B must have increased, or the surface energy of the solid from which the liquid has retreated is greater than it was before wetting (Edser, *Fourth Report on Colloid Chemistry*, B.A.A.S., p. 292). This increase cannot be brought about by adsorption, which would occur only if it reduced the free energy of the surface. It may be that the drop effects some sort of clean-up of the surface, permitting better contact between liquid and solid. Bartell and Cardwell (*Jour. Am. Chem. Soc.* **64**, 494, 1942) measured contact angles of water against the surface of silver prepared in vacuum, and came to the conclusion that hysteresis was in fact associated with the adsorption and desorption of air from the surface of the metal. They point out that measurements of contact angle against soft hydrophobic solids, such as paraffin, which have low surface energies, are generally much more consistent than those against metals. In the case of metals the surface energy is much greater and also more likely to vary in value from point to point. Accurate measurements of contact angles under good conditions have been made but recently and more work is needed to settle the matter.

In technical applications one is usually concerned, not with conditions of static equilibrium, but with more or less rapid movement of liquids over solids (e.g. in washing, painting, flotation, etc.). In such cases there are generally appreciable differences between advancing and receding angles of contact, and it depends on the particular problem whether it is one of the limiting angles of contact or their average value that matters.

For their results to be of use to others, workers should specify carefully their methods of preparing the surfaces and measuring the angles.

(3) *Measurement of contact angles.*

(*a*) *Plate method.* N. K. Adam and others have used this method. A plate *A* of the solid (Fig. 17) is immersed in the liquid and rotated until optical tests show that the liquid and solid meet on one side in a sharp line without curvature. The angle θ is then the angle of contact and this can be measured by a protractor giving the angle which A makes with the horizontal. There are still differences of opinion as to whether this method can be used to determine each of the limiting values for θ ('*Wetting and Detergency*' *Symposium*, p. 48, 1937).

Yarnold measures the contact angle of mercury against glass and steel by measuring the apparent weight of a suspended sphere as it is lowered through, and raised from, the surface of mercury. For

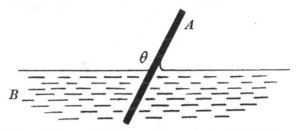

Fig. 17. Plate method of measuring contact angle, θ, between the solid A and the liquid B

finite contact angles there will be one position of the sphere in which the liquid is horizontal right up to the contact line. In this case the upthrust will be wholly due to the volume immersed and not to surface tension. (*Proc. Phys. Soc.* **58**, 120, 1946.)

(*b*) *Contact angles of fibres.* In view of its importance in detergency and allied problems, Adam and Shute (*Wetting and Detergency*, p. 54) have extended the plate method to find the angle of contact of liquids against single fibres, and particularly to examine the contact between a fibre and the interface oɪ two liquids. For the latter purpose one liquid rests on the other in a narrow cell with plane glass sides. The fibre penetrates the interface and the angle is determined when the interface runs right up without curvature to one side of the fibre. Measurements and observations of the contact are made by means of a low-powered microscope.

(*c*) *Contact of bubbles.* The angle of importance in flotation (§4) is that made by air when it displaces water from part of the surface of a mineral. Wark (*loc. cit.*) forms a small bubble at the end of a tube in the liquid and brings the bubble into contact with the mineral surface prepared under the liquid. The profile of the bubble is photographed in order to measure the angle which the gas-liquid interface makes with the solid surface (Fig. 18). Wark finds this angle to be independent of the nature of the gaseous phase, thus indicating that it is not the gas but the water vapour in it that determines the angle.

(*d*) *Contact angle in powders: wetting pressure.* Energy relations at liquid-powder interfaces determine whether suspended particles tend to remain separated, or whether the fluid film between two particles tends to be squeezed out, thus causing the particles to settle into a mass which is difficult to mix effectively with the liquid

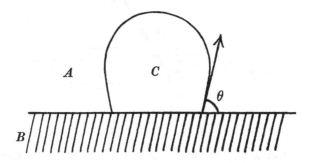

Fig. 18. Contact of bubble and solid. θ is the angle of contact which the bubble C, in the liquid A, makes with the solid B

by stirring. For this and other reasons it is useful to be able to find the contact angle between liquids and finely divided solids. This has been done by an application of the principle of capillary rise.

In the capillary tube [Ch. II, 2 (1)], if the liquid makes an acute angle of contact θ, then when equilibrium is attained we have the relation

$$2\pi r\gamma \cos \theta = \pi r^2 \rho g h$$

or

$$2\gamma \cos \theta / r = \rho g h = p. \tag{1}$$

If then the air pressure within the bore of the capillary tube were increased by an amount p, it would be just sufficient to prevent the liquid from rising in the tube. If p, r and γ in (1) are known, then θ can be found. Bartell and Osterhof (*Colloid Symposium Monograph*, **5**, 113, 1928) have used this idea to find contact angles for powders. A tube is packed with the powder and measurement is made of the pressure which is necessary to prevent the rise through the powder of a liquid which is known to wet the powder completely (i.e. for which $\theta = 0$). This gives, by substitution in (1), the effective r for the powder as packed in the tube. Then, using any other liquid, the measurement of p and γ enables cos θ to be calculated.

This method obviously may be used also to examine the displacement from the powder of one liquid by another, the pressure to prevent displacement enabling the contact angle between the common surface of the liquids and the powder to be determined.

(4) *Surface structure and contact angle.* For substances having large contact angles with water it is found that the apparent contact angles are generally larger for rough or porous surfaces than for smooth ones. Cassie and Baxter (*Nature*, **155**, 21, 1945) discuss the effect of air trapped at a porous surface in increasing the contact angle, and give the relation

$$\cos \theta_2 = f_1 \cos \theta_1 - f_2,$$

where θ_2 is the angle against the porous surface, θ_1 that against the solid material, and f_2/f_1 the ratio of areas of air-water contact to air-solid contact. They give the value $f_2/f_1 = 5$ for duck's feather. This would cause the contact angle against the feather to be greater than 150° though it is less than 90° for the material of the feather. Hence it is the structure of the feathers rather than the material of which they are composed that causes the water to run off.

(5) *Contact angles in liquids: Antonow's rule.* From time to time the validity of Antonow's rule is questioned, and various workers have held that orientated molecules at the interface cause departures from it. Antonow himself (*Jour. Phys. Chem.* **46**, 497, 1942) insists that there are no exceptions under conditions of equilibrium and says that the rule does not apply to conditions when spreading is taking place.

If $\gamma_A - \gamma_B = \gamma_{AB}$ for every case of equilibrium, then the only possible angles for equilibrium in liquids would be 0° or 180°. However, Fox (*Jour. Chem. Phys.* **10**, 623, 1942) has shown that the relation above does not always hold if by γ_A and γ_B are meant the surface tensions of the two liquids in equilibrium with air saturated with their vapours. He found that an air bubble may remain stably at the interface of two liquids, e.g. water and aniline

82 LIQUIDS ON THE SURFACE OF SOLIDS

(Fig. 19). In such a case the relation at the common junction would be

$$\gamma_{AB}=\gamma_A \cos{(\pi-\theta)}+\gamma_B \cos{(\pi-\phi)}$$

and only when one angle becomes 0° and the other 180° will the relation become

$$\gamma_{AB}=\gamma_A-\gamma_B,$$

but then the bubble of air would be surrounded completely by one or other of the liquids.

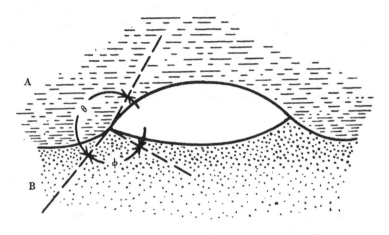

Fig. 19. Air-bubble held at the interface of water and aniline

4. Flotation of Minerals

(1) *Introduction.* It is estimated that the various mining companies now treat a quarter of a million tons of mineral per day by flotation processes. The majority of this ore is of such low grade that the concentration of its valuable portions by any other method would be impracticable or uneconomical. The process is one of separating a powdered mineral into two parts because one constituent of the powder sinks to the bottom of water while the other tends to remain at the water-air surface (cf. the obstinate floating of crystals of boracic acid, although they are denser than water).

The applications of flotation processes have been growing since the beginning of the present century, but it is only comparatively recently that theory has begun to catch up with practice, as the result of work carried out chiefly in America and Australia.

Flotation is briefly discussed in a tract on Spreading because it depends on the same factors as the phenomena of spreading, and particularly on whether water will spread over and completely displace air from contact with the mineral surface or whether air will spread and (partly at least) displace water from the surface.

(2) *Principles of flotation.* The floating on water of a waxed needle and the sinking of a perfectly clean one are well known. Any solid whose density is greater than that of water and whose surface is completely wetted by water will sink. If the particle has a finite angle of contact with water and is not too heavy, it will float

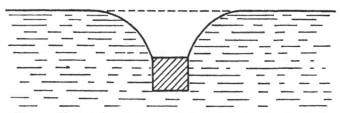

Fig. 20. Unwetted particle supported partly by 'surface tension' and partly by buoyancy

at the surface, supported partly by the vertical component of the surface tension and partly by buoyancy owing to the displacement of water (Fig. 20). For practical purposes it is essential that a particle of mineral which has been immersed shall nevertheless remain at the water-air interface (or float) if agitation of the mixture brings the particle to the surface. This amounts to saying that if the advancing and receding angles of contact differ it is the receding angle which must remain finite. For sufficiently small particles the buoyancy effect is unimportant and only the free energy of the surfaces need be considered. The *decrease* in free energy when air replaces liquid over 1 sq.cm. of the surface of the mineral is given by

$$F = \gamma_A - \gamma_B + \gamma_{AB},$$

where A and B refer to liquid and solid respectively. For equilibrium

$$\gamma_B = \gamma_{AB} + \gamma_A \cos \theta,$$

and substituting for $\gamma_{AB} - \gamma_B$,

$$F = \gamma_A (1 - \cos \theta).$$

Wark (*loc. cit.* p. 51) calls *F* the *tenacity of adhesion* between mineral and air at the air-water interface. The condition of minimum free energy shows that the greater the value of θ the greater the tendency of the particle to be held at the air-liquid surface. Thus in flotation, if a particle of mineral surrounded by liquid is brought by the stirring of the mixture into contact with a bubble in the froth, then the particle will tend to be held by the bubble if its surface is such that the liquid film over it breaks and the liquid retreats from part of the surface of the mineral. The particle is then carried to the top

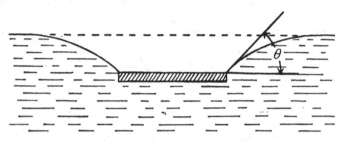

Fig. 21. Floating of large thin sheet of metal on water

of the liquid by the bubble, and so is separated from particles which remain wetted by the liquid. The possibilities of flotation are limited only by the extent to which reagents can be found to modify the surface of the particles so that they may be completely wetted and sink, or only partially wetted and float. Pieces of metal of appreciable weight may be floated at the surface of water if the contact angle is large and the metal is in the form of a thin flat sheet (Fig. 21).

As the size of particles of a mineral is reduced the mass decreases more rapidly than the surface area on which it depends for flotation, so that particles of any density will be held at the interface so long as they are small enough and have a finite angle of contact with the liquid. In practice there will clearly be an optimum size for particles (actually 50–150 microns), a finer grain merely necessitating a larger surface to float a given amount and also a greater expenditure of reagents to produce the necessary 'hydrophobic' or 'hydrophilic' character over the surfaces.

(3) *Flotation reagents.* In practice various substances are added to the water to effect the separation of minerals. From their

functions these reagents are classified as frothers, collectors, activators, or depressants, though in some instances a frother may also act as a collector (Wark, *Principles of Flotation*, 1938; *Jour. Phys. Chem.* **40**, 799, 1936; *Proc. Aust. Inst. Min. and Met.* **90**, 1, 1933, etc., where references to recent technical literature are given).

(*a*) *Frothers.* As the name suggests, these are substances added to the water so that the agitation of the mixture will produce a more or less stable foam, thus giving a large water-air surface to carry the particles. Edser (*Fourth Report on Colloid Chemistry*, B.A.A.S.) analysed the factors on which the stability of foam depends. When a bubble is deformed its persistence depends largely on the difference between its dynamic and static surface tensions (Ch. II, 2). The extension of the film reduces the concentration of adsorbed molecules at that part where the extension occurs, thus raising the surface tension. A bubble persists if these adjustments occur readily, deformations being met by variations in surface tension instead of immediate rupture of the film.

In flotation the adhering mineral particles themselves play an important part in preventing the coalescence and collapse of the foam. In general, frothers should not produce too large a reduction in surface tension, since this lowers the adhesion of the particle at the water-air interface. Wark cites pine oil as a frother which, in the concentration necessary, reduces the surface tension of water by only $\frac{1}{4}\%$.

(*b*) *Collectors.* These are substances which are adsorbed at the surface of particles of the mineral to be floated, probably as a fairly complete monomolecular layer. Collectors in general have molecules possessing an active and an inactive portion. The active ends are fixed at the mineral surface and the inactive groups, presented outward, give the necessary hydrophobic characteristics which permit of a finite angle of contact and so of flotation in the froth. For sulphide minerals organic compounds known as xanthates are most generally used, though theoretically any substances which will be adsorbed at the surface of the particles and produce a finite contact angle may be used.

(*c*) *Activators.* Substances which will react with molecules at the surface of the mineral producing a surface to which the collector may be adsorbed are called activators. For instance, certain sulphides (e.g. ZnS) are not floated when ethyl xanthate is used as a collector, but if a copper salt is first added to the mixture the xanthate will then act and cause the mineral to adhere to the foam. The particles become coated with a layer, perhaps only unimolecular, of CuS to which the xanthate is adsorbed.

Activation may be produced by the presence of 1 milligram per litre of $CuSO_4$, an amount that may be provided by traces of copper present in the mineral.

(d) *Depressants*. These are substances (usually alkalis or cyanides) which act selectively in preventing a mineral from being floated. They maintain on the particles a surface such that water spreads over it and wets it, causing it to sink. In general, depressants are used to prevent the adsorption of the collector.

By the use of collectors, depressants and activators it is now possible in many cases to differentiate between the sulphides and effect a practically complete separation of the constituents of the minerals.

More recent work has not changed the general principles of flotation. A recent application of interest was the successful separation of a 99% pure sample of ergot from rye by a flotation technique developed by Sutherland and Miss Plante (*Jour. Phys. Chem.* **48**, 203, 1944). The valuable ergot replaces some of the grains in the rye. Difference in density is not sufficient for a separation by gravitational means but by treatment of the grain with an oil emulsion the rye grains were floated leaving an almost pure sample of ergot. Grains of rye are probably the most massive 'particles' which have been separated by a flotation process.

5. Detergent Action

The removal of an oil film from fibres of wool is important in industry, and the washing of fabrics generally involves the removal of dirt which is held to the fibres by a layer of grease. The natures of grease, dirt and fibre all enter into the problem of detergency, and it is now realized that there is not likely to be any simple theory of detergent action or any single property of the detergent solution in terms of which its efficiency can be expressed. Adam, Robinson and others ('*Wetting and Detergency*' *Symposium*, 1937) have done much to show what are the fundamental factors in detergency.

With the camera Adam has shown that the oil film is loosened from the fibre and then forms droplets which are carried away by the solution owing to mechanical agitation.

When the dirt and grease are removed from the fibres and exist as an emulsion in the water they must be removed without breaking down the emulsion and redepositing the dirt on the fabric. This break-down of the emulsion can occur if the rinsing water contains calcium or if the change in temperature or concentration of detergent is too sudden (cf. Tomlinson, *Manufacturing Chemist*, **15**, 159, 1944).

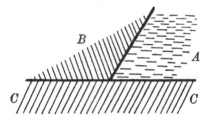

Fig. 22. Liquid *A* tends to move to the left displacing liquid *B*
from contact with the solid *C*

The initial replacement of oil film by detergent solution depends on the contact angle between the oil-solution interface and the fibre. So long as the contact angle (in the solution *A*, Fig. 22) is acute the solution will tend to advance, displacing the oil *B* from the fibre *C* (cf. capillary rise when the contact angle is acute and depression when it is obtuse). Adam (*loc. cit.*) finds that for efficient detergency the contact angle should be zero, and this would mean energy conditions favouring the complete wetting of the fibre by the solution and hence the removal of the oil layer. If the energy conditions are favourable for the preferential spreading, then the mechanical agitation of the fibres of course assists the spreading, by compensating to some extent for the immobility of the molecules in the surface of the fibre (Ch. IV, 4).

Recent advances in detergency have been chiefly in the production of chemical substances which, even when used with hard water, do not form insoluble precipitates to be deposited on the fabric, nor form a scum on the water, as do the stearates of ordinary soaps.

Tomlinson (*loc. cit.* p. 198) gives a summary, with references, of work on the solvent action of detergent solutions. At concentrations higher than that at which micelles form, organic substances dissolve in soap solutions, forming true solutions, not emulsions. The hydrophobic chains in the micelle function as a liquid hydrocarbon and the molecules dissolve in this. When the detergent is present in fairly high concentration, as in washing the hands with soap, this direct solution, as well as true detergent action, probably comes into play.

6. Other Technical Applications: Wetting Agents

There are many other technical processes in which the stable wetting of a solid by a liquid is a fundamental factor. The general principles involved have been sufficiently discussed. An account of recent research work on this subject with references to the technical literature is given in the Report of the *Symposium on 'Wetting and Detergency'* to which reference has already been made.

(1) *Wetting of metals by liquid metals.* In the tinning and soldering of metals the wetting of the solid metal is promoted by means of fluxes. From the general conditions of spreading we should expect the function of the flux to be firstly to clean the surface, raising γ_B, and probably also to lower the interfacial tension γ_{AB}. Of course, spreading occurs more readily if γ_A, the surface tension of the solder or liquid metal, is lowered. However, as is pointed out by Daniels and Macnaughtan ('*Wetting and Detergency*' *Symposium*, p. 77), where it is desired that the solder should penetrate into joints (e.g. in sealing cans), a high surface tension is desirable in the solder. So long as the solder wets the metal, the higher its surface tension the farther it will be drawn into crevices.

(2) *Wetting agents.* In the application of paints, varnishes, dyes and even horticultural sprays the results depend on the stable wetting of a solid by the liquid. Much attention is being devoted to the production of wetting agents which may be incorporated in the liquids in order to increase adhesion between liquid and solid and so stabilize the layer of liquid once it is spread or assist it in spreading. The essential property of the molecules of wetting agents is the possession of a polar and a non-polar group rendering

them 'amphipathic'. Modern wetting agents may consist of molecules possessing several active groups, and their production and properties are matters of chemistry. The physical principle involved is the same as in the case of 'active' substances increasing the adhesion of a lubricant (§2).

Dean ('*Wetting and Detergency*' *Symposium*, p. 25) gives a survey of the chemistry of wetting agents and also references to the patent literature where they are described.

INDEX

Accuracy of measurement, 10
Acids, fatty, 37
— — in lubricants, 73–6
— — in non-spreading paraffins, 61–3
— — on water, 53–5
— promote spreading of water on mercury, 43–4
Activators (flotation), 85–6
'Active' substances, in lubricants, 73–4
— in non-spreading paraffins, 61–3
— in spreading, Chs. V, VI
— produced in paraffin by irradiation, 66–8
Adhesion, of liquids and solids, 38–9
— tenacity of, 84
— tension, 39
Adsorption, at boundaries, 3–6
— attempts to show in pure liquids, 5
— of gases on mercury, 29–31
Alkalis, electrically forced spreading, 49
— retard spreading on mercury, 42
Alloys, 25
Angle of contact, 38–9
— and surface structure, 81
— hysteresis of, 77–8
— in detergent action, 87
— in liquids, 81
— methods of measuring, 78
— of bubbles, 79
— of fibres, 79, 87
— of powders, 80–1
Ångström, unit of length and area, 54
Aniline, spreading on water, 68–9
Antonow's rule, 50, 81

Benzene, on water, 68
— surface tension of, 13
'Body' in oils, 75
Bond's flowing sheet method, 20
Bubbles, contact angle of, 79–80

Calcium ions at interfaces, 64
Calming of waves by oil and rain, 69–70
'Capillary constant', 13
Cohesion of liquids, 38–9
Collectors (flotation), 85

Complexes in monolayers, 64
Contact angle, see Angle of contact
Contamination, effect on spreading, 42
Curvature of film, 10–11

Depressants (flotation), 86
Detergency, 86–8
Dilute solutions, minima in γ-c curve for, 7
Double layer, 49
'Duplex' films, 59–60
Dynamic and static methods for γ, 12

Electric current controls spreading, 48
Electro-capillary effects and spreading, 49–50
Energy, conditions for spreading, 34–6
— surface, 1
Evaporation, influence of surface layers on, 70

Fatty acids, see Acids, fatty
Films, condensed, 58
— expanded, 58
— gaseous, 57–8
— homogeneity of, 60–1
— monomolecular on water, 55
— multilayer, on solids, 65–6
Flotation, 82–6
— reagents, 84–5
Free energy, and adsorption, 24
— change, and flotation, 83
— relation and surface tension, 3
Frothers (flotation), 85

Gallium, 32
Gases, on the surface of mercury, 29
— effect of, on surface tension, 29
— persistence of monolayer of, 30
Gravitational forces, and spreading, 41

Immobility of surface, effect on spreading, 39–40
Interfacial films and monolayers, nature of, 64
— tension, 22–3
— — of mercury and pure water, 50
'Invisible' glass, 65–6
Ions, effect on spreading, 42–4

Ions, metallic, *see* Metallic ions
Isothermal expansion of gaseous films, 57

Langmuir-Adam balance, 55
Large drop, correction for finite size, 16–18
— method applied to water, 17
— surface tension by measurement of, 14–15
Liquids, non-spreading, 52–3
— on the surface of solids, Ch. VII
Long-range forces, between molecules, 41
— in lubrication, 75
Lubrication, 73
— boundary, 74–6
— complete, 73–4

Maxima and minima in γ-c curves for dilute solutions, 7
Mechanism of spreading, on liquids, 36–8
— on solids, 38
Meniscus corrections in manometers, 27
Mercury, 25–8
— discordant values for surface tension, 25
— surface tension in gases and vacuum, 26
— temperature coefficient of surface tension, 28
Metallic ions, effect at interface, 64
Metals, high surface energy of, 24–5
— surface tension of a liquid, Ch. III
Micelle, 9
Migration, molecular, 41
Molecular size from observations on spreading, 54–5
Monolayer, complexes in, 64
— durability of, 76
— visible limit to, 68

Non-spreading liquids, 52–3

Oil, calming effect on waves, 69–70
'Oiliness', 75
Overturning of molecules in films, 65

Paraffins on water, 53
— spreading caused, by ultra-violet light, 66–8
— — by active substances, 61–3

Polarization and interfacial tension, 51
'Pressure' in films, 57–60
Protein layers, 65

Radioactive isotopes to record wear, 76

Sessile drop, *see* Large drop
Sharp, 4
Sodium oleate, 4
Solids, liquids on surface of, Ch. VII
— molecular migration on, 41
Spreading, and electro-capillary effects, 49–50
— chemical and physical theories of, 33
— energy conditions for, 34–6
— coefficient, 34–6
— controlled by electric current, 48
— controlled by nature of ions, 45–7
— general conditions of, Ch. IV
— lack of complete theory of, 33–4
— of liquid metals, 88
— of liquids on mercury, Ch. V
— on liquids, mechanism of, 38–41
— on solids, mechanism of, 38–41
— on water, Ch. VI
— to a limited area, 43–4
— 'vapour pressure', theory of, 40
Stability of foam, 85
Static and dynamic methods for γ, 12
Surface energy, and spreading, 34–6
— and surface tension, 1
— origin of, 2
Surface layer, composition of, 3–4
— evaporation through, 70
Surface migration on solids, 41
Surface potential, 61
Surface tension, curvature and pressure, 10–11, 16
— and critical temperature, 12
— measurement of small differences in, 22
— methods of measuring, 11–22
— — adhesion of rings, 20
— — bubble pressure, 14
— — capillary rise, 12–13
— — drop-weight, 13–14
— — flowing sheet (Bond), 20
— — large drop, 14–15
— — pendent drop, 21
— of liquid metals, 24–5
— of mercury, 25–31
— — fall with time, 4–5

Surface tension of mercury, temperature coefficient of, 28
— unit of, 1

Temperature coefficient, of γ for mercury, 28
— positive, 3
Time-lag in reaching equilibrium, 4

Ultra-violet light, action of on paraffin, 66–8

Vapour pressure, in films, 57–8
— theory of spreading, 40

Water, as a solution of ions, 51
— spreading on mercury, 43–7
— surface tension of, 12–21
Waves, action of oil on, 69–70
Wetting agents, 88–9
— of metals, 88
— pressure in powders, 80–1

Printed in the United States
By Bookmasters